OLD SOLDIER SAHIB

Private
FRANK RICHARDS
D.C.M., M.M.
Late Of The Second Battalion,
Royal Welch Fusiliers

WITH AN INTRODUCTION BY
ROBERT GRAVES

LIBRARY OF WALES

Parthian, Cardigan SA43 1ED
www.parthianbooks.com
The Library of Wales is a Welsh Government initiative which
highlights and celebrates Wales' literary heritage in the English language.
Published with the financial support of the Welsh Books Council.
www.thelibraryofwales.com
Series Editor: Dai Smith
First published in 1936
© Estate of Frank Richards
Library of Wales edition published 2016
ISBN: 9781910901205
Cover: Regiment of the 33rd Punjabis
Culture Club/Hulton
Archive/Getty Images
Typeset by Elaine Sharples
Printed and bound by Gwasg Gomer, Llandysul, Wales

To the Prayer Wallah
in the hope that this meets his eye

CONTENTS

FOREWORD

The charm of this book is obvious enough, but to recognize its historical importance one must first consider the old-fashioned infantry soldier who built up the British Empire in the seventeenth, eighteenth and nineteenth centuries. He was known on the one hand for his foul mouth, his love of drink and prostitutes, his irreligion, his rowdiness and his ignorance; on the other for his courage, his endurance, his loyalty and his skill with fusil and pike, or with rifle and bayonet. Wellington referred to the troops who served under him in the Peninsular war as "the scum of the land," but they won him a dukedom by driving the French out of Spain in a series of extraordinarily severe engagements, and finally defeated Napoleon at Waterloo. It was only after Waterloo that a generic name was found for the British soldier. It was adopted from the sample filled-out form shown to the troops to help them record the details of their service correctly: "I, Private Thomas Atkins, of His Majesty's Twenty-Third Regiment of Foot." There is a legend that Wellington himself supplied this name and regiment in commemoration of a soldier who had come under his notice in Spain for gallantry in the field. A new name for a familiar type.

The army that fought and swore and drank in Spain was composed of much the same sort of heroic scum as had fought and swore and drunk in the Low Countries under Marlborough a hundred years before; and it remained much the same throughout the nineteenth century, and until the Great War. It was still Thomas Atkins who in 1914 fought under Sir John French on the Marne and the Aisne, and who then as a last proof of his courage, his discipline and his marksmanship saved the Channel Ports at the First Battle of Ypres and so faded away into history.

Thomas Atkins was temporarily succeeded by the citizen soldier, and at the conclusion of the War by a new type of professional soldier, a man with a far higher standard of education, far greater sobriety and a strong mechanical bent. Beer and the rifle ceased to be the two main ingredients of Army life. Already towards the end of the War it was being jokingly said in my battalion (first, I believe, by the author of this book) that with the formation of all the new specialist groups – machine-gunners, Lewis-gunners, trench-mortar men, bomb-throwers, rifle-grenadiers, gas-specialists – the ordinary rifle-and-bayonet man would soon be an out-of-date survival who would parade at the head of the battalion on Church Parade in company with the Regimental Goat. It has not yet quite come to this, but the mechanization of the Army is progressing steadily and experts in modern war allow the ordinary infantry-man merely the status of "mopper-up" after an offensive launched with tanks, armoured cars, gas, artillery, and aerial bombing and machine-gunning.

Thomas Atkins is gone. What records remain to tell what sort of a man he really was? Few, and most of those misleading and contradictory. Some of his officers have written about him, in a gentlemanly but distant style, and he appears, more or less in caricature, in novels and plays of the eighteenth and early nineteenth centuries; and there are one or two miserably sentimentalized accounts of him by Victorian authors – for example, Dickens' *Private Richard Doubledick*. Perhaps the best miniature portrait of him that survives is contained in the first stanza of Sir Francis Doyle's *Private of the Buffs*. According to the note which accompanies the poem, the private was No. 2051 Private John Moyse, and he and some Sikhs "having remained behind with the grog-carts," during an incident in the 1860 campaign in China, "fell into the hands of the Chinese." They were ordered to kow-tow before a Tartar governor. The Sikhs obeyed, but "Moyse refused to prostrate himself before any Chinaman alive and was immediately knocked upon the head and his body thrown on a dunghill."

Last night, among his fellow roughs,
  He jested, quaffed and swore,
A drunken Private of the Buffs
  Who never looked before;
Today, beneath the foeman's frown
  He stands in ELGIN's place,
Ambassador from Britain's Crown
  And type of all her race.

Poor, reckless, rude, low-born, untaught...

The poem deteriorates here, but the point has been well made. Elgin, by the way, was British Envoy to China at the time (and afterwards Viceroy of India) and the Buffs are the East Kent Regiment, so called from the buff-facings on their uniform.

But when did Thomas Atkins ever publish any intimate statement about himself? Or how could he? He was an unlettered man and even if he had a gift for story-telling and could find an amanuensis, the sort of story that he had to tell would not have been considered polite material for publication. Occasionally a sergeant or sergeant-major wrote an account of his own part in some campaign – one Sergeant Lamb, for example, of the Twenty-Third Foot (the Royal Welch Fusiliers) has left a valuable record of his service in the American War of Independence. But the questions that we should like to ask about the way the troops lived in peace and war are never answered in such works; and the language is usually more studied and frigid, even, than the language of officer-historians.

Rudyard Kipling's *Soldiers Three* was a literary surprise. He seemed somehow to have won the authority of Thomas Atkins to make on his behalf the first public revelation of barrack-room life in the East, with all his private joys and sorrows and sins laid out as for kit-inspection. A good deal of what Kipling wrote was interesting and accurate; a good deal was necessarily in the vein that he afterwards exploited in "Thy Servant a Dog" –

experimental intuition of the feelings of a poor dumb friend; a good deal was fantasia. His Privates Ortheris, Learoyd and Mulvaney are, we feel, far closer to the truth than Dickens's Private Doubledick, but they are clearly not the real Thomas Atkins. It was impossible that they should be. Kipling was, in the first place, a civilian and never lived a barrack-room life himself; next, he was a journalist with the journalist's temptation to improve on a good story; and, lastly, he wrote the book at a time when nearly all the best stories that he must have picked up in conversation with old soldiers and quite all the swear-words that embellished them were still considered unprintable. What "ruddy" and "blushing" and "perisher" were to the real swear-words, Kipling's amended stories were to the originals.

Now, unexpectedly, Private Frank Richards, who happens to belong to the same regiment as did Sergeant Lamb and the original Private Thomas Atkins, has supplied the very document that has been so long missing from the archives. It is an old-fashioned British soldier's own record of service, written in his own language, unimproved and uncensored: "I, Private Thomas Atkins, of His Majesty's Twenty-Third Regiment of Foot ..." The words "His Majesty's" are not idle. "Our Royal Family was always popular with the rank and file," Richards reports.

The word "uncensored" reminds me of the time that I first became aware of Richards' talents as a writer: it was in trenches, two days after the Battle of Loos, in September, 1915. He was a signaller in my company, and we had both been lucky not to be killed: the Battalion had lost heavily in an attack against good German troops, posted behind uncut barbed wire. There had also been a number of gas-casualties from our own gas, which was being used for the first time: it was altogether a bad show. I was now censoring the company letters, as usual a dull task: "This comes leaving me in the pink which I hope it finds you. We are having a bit of rain now. I expect you will have read in the papers of this latest do. I have lost a few good pals but was lucky myself. Fags are always welcome, and also socks." Seldom any more than that, and signed "ever your loving husband," or "ever your respectful son." Then I came on a long letter addressed to someone in Wales and giving a detailed, accurate and grimly joking account of what had really happened – the uncut wire, the bungling of the gas, the casualties, the strange behaviour of the Scottish troops on our flank, the readiness of our own men to continue the suicidal attack, had not Major Charlie Owen, the Adjutant, in temporary command, sworn angry oaths and called it off.

I handed the letter to a regular officer (one of the very few left) who was sharing the dug-out with me, "Can I pass this?" I asked.

He read it through. "The man ought to be crimed, really. But every word is true, and the official communiqué will be all lies, and people at home ought to know what really happened. You know this Richards, of course? – a good man. The best signaller we've got, and the oldest soldier in the company – fifteen years' service. Here, I'll pass it myself." He scribbled his initials on the envelope.

I was in the Battalion with Richards in 1915, in 1916 (when I was wounded during our capture of High Wood), and early in 1917; after which I was invalided home for good. Richards, who had been with the first troops who arrived in France after the declaration of War, brought off a twenty-thousand to one chance: he stuck it out until the Armistice, never missing a battle, never getting wounded, never applying for promotion or a transfer, serving under a dozen successive commanding officers, seeing the Battalion constantly smashed up and constantly reorganized. He was at the battle of Le Cateau in August, 1914, and again at Le Cateau when the German advance ebbed back there in November, 1918. His Distinguished Conduct Medal was won at a time when few were given, because of the pension that went with it. The Military Medal, which carried no pension, had been instituted as the commoner decoration; Richards won that too. I admired Richards – tall, tough, resourceful, Welsh, and the company humourist – and was as friendly with him as a very young officer was permitted to be with a very old soldier: indeed, the burden of keeping up the social conventions fell chiefly on Richards.

In civil life Richards was a timber-man in a South Wales coal-mine, but the War left him with a weakness that prevented him from continuing work down the pit, and he had to take up clerical work, occasionally getting a job in a Labour Exchange. Three years ago in an interval of unemployment he wrote out the story of his war-experiences in France and Belgium, and sent it to me one day as a surprise, asking whether it was worth anything. I told him, yes, it was worth a lot, and that I would do my best to find a publisher for it, but that as it would be classed as a War-book, and as War-books were no longer in fashion ... However, it turned out better than I had expected. *Old Soldiers Never Die* was widely read, and as a book about the Army rather than about the War.

I then urged him to write out the story of his pre-War soldiering in India and Burmah, to make the record complete. This was something that would interest people even more than the other book. He said that he would try, but that he found pen-pushing a wearisome occupation and had enough of it, filling out forms all day long at the Labour Exchange. I persisted and eventually he sent me the task completed, remarking that he thought very little of it, and that if I was of the opinion that it should be destroyed I should frankly say so; but, if not, would I kindly (as I had done with the previous book) glance my eye over the punctuation and the paragraphing – he had never been good at that sort of thing. This I was glad to do for him. I also asked him a number of questions about barrack-room life in India, and about the authenticity of certain odd stories I had heard, twenty years before, from old soldiers in the Battalion. His answers he has incorporated in the story.

Richards' literary style witnesses to his long training as an Army signaller. Each anecdote or descriptive passage is like a private message flagged out or sent by helio to a friend at a neighbouring signal-station with the formal "AAA" at the close.

His reported dialogues and monologues read like authentic speech memorized for sending. Certain provincialisms occur – "enough of money," "the both of us," "better than what I could have done" – and frequent zoölogical metaphors, but the style is lucid, economical and never in the least pretentious. The end of the adventures of Gerald the gentleman-ranker, where Gerald pays for the beer of fifty time-expired men and for their bread and cheese, "and the pickles too," illustrates Richards' method. A less gifted writer would have mentioned only the beer, or perhaps gone as far as the bread and cheese: but "the pickles too" are to this story what the extra twopence given to the innkeeper is to the Gospel story of the Good Samaritan. Similarly with Richards' account of the fight between the two silver-bellied eagles on a rock high up in the Himalayas.

> When we reached the rock we both gazed in wonder at the great splashes of blood all over it, and asked each other how the birds could have flown at all after losing so much blood; what struck us also was the small quantity of feathers that they had lost, considering how viciously they had used their wings.

Many writers could not have resisted the temptation to make the battle-ground as littered with feathers as a poulterer's shop in Christmas week. Many writers, also, would have failed to carry on the story of Bern the murderer beyond the point where he was sentenced to death. But Richards has no fear of anti-climax, because the importance of the story lies less for him in the simple tale of a man sent mad by wounded pride and affection, than in the relation of individual tragedy to the whole disinterested military setting: more important than the fact of murder, to the troops at Kilana, was the keeping up of race prestige during the hanging. Richards' AAA is only signaled after the Bishop has finally sanctified the grave amid the ribald remarks of the men who witnessed the ceremony.

His success in conveying the meaning of Indian summer heat on the Plains by a quiet enumeration of its physical effects on the troops is remarkable. Kipling's method was to introduce the element of hysteria (as in his story of the man who ran amuck on the parade-ground with a rifle) leaving the physical rashes and blisters to the imagination. Richards' only hint of any weakening of spiritual resistance is his story of H company's hoodoo, the crow in the cemetery; but frayed nerves are sufficiently indicated by the midnight cry of "Cinch, you black bastard," and the running kick at the ribs of the exhausted punkah-boy.

The simplicity of Richards' outlook is based on a moral sureness which seems largely the result of his upbringing. The South Welsh miner has a very limited moral code, but sticks to it; and it tallies well with the Army code. The prime virtues both in pit and barrack-room are courage and openness, generous loyalty to one's friends, and honesty. "Thou shalt not kill, thou shalt not steal, thou shalt not commit adultery" only make sense in the context of these virtues. "Thou shalt not kill – in a mean or cowardly way." "Thou shalt not steal – from a friend." "Thou shalt not commit adultery – and be a respectable chapel-goer." Ostentation and inefficiency seldom deceive the old soldier or the old miner. Note, for example, Richards' verdict on Sir Hall Caine's novels, and on the turnouts in the parade of Rajahs at Agra.

It is unlikely that anyone will quarrel seriously with the broader passages in this book. Richards never sniggers in the perverted modern fashion; he manages his anecdotes as the more vigorous eighteenth-century novelists did, never letting disgust get the upper hand of laughter in his reader. He writes in the same language as he talks, with the natural eloquence of the South Wales miner. He has complained to me of the difficulty of knowing what one can say in print without getting into trouble, for he always expresses himself freely in conversation and "what would shock a soldier would cause a packman to drop his pack." As for his opinions about the proper treatment of orientals, he would not be the authentic Thomas Atkins if he held others. ("Let dusky

Indians whine and kneel," as the Private of the Buffs had held.) The opinions that Richards' predecessor, the old Bacon-wallah at Meerut, held about natives were still more severe; that they were not unrepresentative of the Army in India during the Mutiny can be judged by the bloodthirsty letters that young Lieutenant Roberts, V.C., afterwards Earl Roberts of Kandahar, wrote home at the time to his family. Since Richards' day there has been a further decline in severity. What has been won by the sword is freely ceded by the pen. Indians wear boots when and where they like and no punch on the spleen follows. Curzonism has conquered. It should, however, be noticed that Richards and his comrades showed a strong sense of justice in their dealings with Indians: they did not maltreat them unless they offended first. The old soldier (and a man did not remain a young soldier for long in India) felt the responsibility of his title "Sahib" and, as "ambassador from England's Crown," seldom behaved like a bully. A man like Private Crickett who cheated the bungalow-servants of their pay was the exception, not the rule; the story of Private Robb's murder of the Chinaman was incomplete, for Richards, without the moral sequel of retribution; and Richards is careful to point out that "The Soaker," who robbed the Ragwallah at Jhansi, belonged to a regiment not his own.

I do not think that there is anything further for me to say, except that wherever I have been able to check the story it has proved correct; and that I hope that its popularity will encourage the publishers to issue *Old Soldiers Never Die* as well – a most moving story and a perfect companion-picture to this.

ROBERT GRAVES

OLD SOLDIER SAHIB

Private
FRANK RICHARDS
D.C.M., M.M.
Late Of The Second Battalion,
Royal Welch Fusiliers

ENLISTMENT

I have some recollections of my father and mother but will commence my story in the year 1893, when I was left an orphan at the age of nine and adopted by an uncle and aunt who were living at Blaina, Monmouthshire. My Uncle, who was Welsh, was a twin-brother to my mother; my Aunt was also Welsh, and at the time of my adoption they had four children of their own, two boys and two girls. David was three years older than me, Evan was about the same, the girls were both younger. As the years rolled on more children arrived, but my Uncle and Aunt treated me exactly the same in every way as they did their own children: no boy could have had better parents than what they were to me. I was quite happy with my cousins and we always shielded one another in any scrape we were in; even today we still regard ourselves as brothers and sisters.

Blaina was a busy industrial town, with seven coal-pits, a large tinplate works, blast-furnaces and coke-ovens – all working regularly. In those days if a man or a boy was dissatisfied with his job he could always leave it and immediately get another. In fact, there was such a shortage of labour that one of the bosses of the blast-furnaces, situated just off the main road, used to stand on the road the greater part of the day, stopping tramps as they came along and begging them to sign on and start work at once. My Uncle was a roller in the tinplate works and earned very good money. He was fond of his pint but always looked after his home. He had never been to school in his life and could not read or write, not even so much as his own name. But he was shrewd about money and figures and could work out sums in his head quicker than we children could on paper. Nothing pleased him better than when my Aunt read something out of the paper to him which interested him: he would get it by heart and be able to repeat it very nearly word for word some days later. He could speak a little Welsh, but it did not come natural to him to do so.

My Aunt spoke Welsh as well as she did English, or better, and she could also read and write both languages fluently. She did her best to teach us children Welsh, but the little I learned as a boy I soon forgot in after life. My Uncle regretted his own lack of education and was always drumming in our ears the importance of attending school and never missing a lesson: the only time he ever severely chastised anyone of us was when he discovered we had been playing truant. There was a school-attendance officer in the district, but he did not have the power at this time of summonsing the parents of a child who was absent from school: all he could do was to report the fact of the child's absence to the parents and leave them to deal with the matter. Evan and I detested school; we were always thinking of the time when we would be twelve years of age and could begin real work. Between the ages of ten and twelve we must have been absent from school more times than we attended it; Evan and I used to keep count of the number of hidings we had from my Uncle and the number of canings we had from the schoolmaster during that period, but I for my part have forgotten the final score.

I can't remember my Uncle ever going to Church or Chapel; I often heard him say that the men he associated with in the pub had a more genuine Christian spirit than the men who attended Church or Chapel. My Aunt was a little religious, however, and often attended Chapel on a Sunday morning, taking us children along. After dinner she would send us to Sunday School, which I also detested. Only twice did I go there and afterwards invariably played truant, going birdnesting or roaming the mountains; and came home to my tea with a far better appetite than what I would have had if I had sat through Bible lessons. My Aunt often said that I was fast going to the Devil and when she found out, just before I began work, that I had ceased to say my nightly prayers she started to cry and said that she was afraid I was already in his clutches.

I left school and began work on the very day I was twelve. My first job was door-boy in a colliery. I had to be down the mine before seven in the morning and finished in the evening at five o'clock. My job was an easy one. I had to open and close one of the ventilation-doors for the hauliers and their ponies to pass through with their full or empty trams. I was paid seven shillings and sixpence a week. I worked two months as a door-boy and then got a job in the tin-works. This also was an easy job: with a bucket of grease and a brush I greased the cold rolls, as they were called, to prevent them from getting heated. The majority of the tin-workers did eight-hour shifts, but the boys and young women who worked at the cold rolls did twelve hours. My wages were now nine shillings a week. I never became an expert at this job and somehow managed to get more grease over my clothes in one day than what I could put on the rolls in a week. I had soon more or less ruined our kitchen chairs: when I returned from work and sat in one of them I found that I could not rise without bringing it up also, stuck fast to my backside. This job did not suit me at all and it also affected my health. More for the sake of my health than the furniture my Uncle and Aunt decided that I must go underground again.

A week or two later I was back in the pit working for a man in a stall at the coal-face, who paid me a weekly wage of eleven shillings, which was considered a high wage for a boy of twelve. The stalls were driven to left or right of the main headings. Although in the same seam of coal, some stalls were better for working in than others; some had a good roof overhead and coal easy to cut, while others had a poor roof and difficult coal. I had to work very hard at my new job, but I did not mind that. My buttie was an excellent man in every way and although sometimes he swore ferociously, he never swore at me or even grumbled at my work during the six years I worked with him.

The pit I worked in was one of five owned by a private company. There was a head-manager who visited each of the five in turn. He was a deeply religious man who attended Chapel regularly; and the majority of the over-men, foremen and staff of the pits followed his example, to keep on the right side of him. It was noticeable that most of the good places in the pit were worked by men who were deeply religious, so men who had hard places found it advantageous to turn deeply religious too, and attend the Chapel where the manager worshipped. He noticed their presence and as soon as their bad stalls had finished they were sent to turn new stalls where the roof was good and the coal easy to cut. Stalls were rarely driven more than sixty yards; they were then cut by headings which were driven for that purpose. In one stall where we worked my buttie and I could always tell when the manager was down the pit and approaching us through the stalls on our left: because the man in the stall immediately on our left, as soon as he saw the manager with his staff approaching, would immediately strike up a hymn. By the time the manager reached his place he was working and singing like a demon. But although this man attended the correct Chapel and although the manager usually stopped for a few minutes' earnest chat with him when he came by on his rounds, he never got the main heading which he was angling for, where more money could be made than in a stall. I never cared much for this hymn-singer; he was always grumbling at the boy who worked with him and calling him a lazy young hound. After working alongside him for nine months I came to realize how lucky I was to be working with a man who made no pretence at being a Christian but succeeded better than many of his deeply religious mates.

It was the custom at the dinner hour for a number of men and boys to collect together underground in one road. I enjoyed myself at these dinner hours, especially when a roadman who cleared the main headings of little falls of rubbish was present. He was an old soldier who had served in different countries and his wonderful yarns of India so enthralled me that I vowed that when I was old enough I would be a soldier and see this wonderful place which, he said, was a land of milk and honey. He was always telling us what a damned fool he had been to leave the Army, and that if he had his time over again he would enlist at once and not take his discharge until it was forced on him.

I was far healthier working underground than I had been in the tin-works. The air was quite good in the workings. The head-manager had the main airways kept in excellent condition, and saw that the roof and timber overhead in the roads were high enough to allow a pony to walk along without roughing his back. The miners were keen on sport, especially rugby football. Pick-up games were frequent at the pitheads and the play though friendly was extremely rough. Blaina had always had a good rugger-team. About 1900 soccer was introduced, but rugger will always remain the miners' favourite game, I believe. In agreement with the owners all miners took a holiday on the first Monday in every month: this was called "Mabon's Monday" after the principal leader of the South Wales miners. Many matches that had been made for whippet-racing and rabbit-coursing came off on this day in the fields on the mountain-side. There was cock-fighting too, which was illegal and had to be arranged carefully with scouts put out to signal the arrival of the police. An occasional knuckle-fight also came off on the top of the mountain. I remember witnessing one bout from which much had been expected. One of the principals was a dashing fighter and the other was said to be a skilful man at fighting on the retreat. Both lived up to their reputations on that day. The fight ended in a draw seven miles from the spot where it had started. It was not customary to fight in a pitched ring, and the retreating fighter kept his face to his foe throughout the battle: he was a man so clever on his feet that he could have easily walked backwards, without stumbling once, from John O' Groats to Land's End. My cousins David and Evan were keen on boxing, and so was I; we were always practising together in the cellar of our house. David was too strong for us, but it was good practice.

I had well passed my fourteenth birthday when the whole of the miners in South Wales came out on strike. This was in 1898 and the strike lasted six months and, like all other strikes before or since, ended up in a victory for the coal-owners. A year or two before this the tin-works had closed down, and my Uncle had been working at various outside jobs for hardly a third of the money that he had been earning in the tin-works. He had never worked underground in his life and always said that he never would; and he never did. He was a very big man, weighing between fifteen and sixteen stone and as strong as an ox. A month after the strike began he got a job, rolling in some tinplate works about twenty miles from Blaina. He lodged there during the week and came home for week-ends. Luckily for the family, he was now earning sufficient money to provide for the whole household, and we were never short of anything during the strike. But the strike threw David out of work and he enlisted in the South Wales Borderers. My Aunt was terribly cut up about it. She wrote and told him that if he did not like the Army he was to let her know at once and she would find the money to buy his discharge. He wrote back saying that he was as happy as a trout and that he intended to complete the term of service that he had enlisted for, which was seven years with the Colours and five on the Reserve. I had a glorious time during the strike. Most of it I spent watching the sport on the mountain-side or playing marbles. I was considered a bit of an expert at marbles, and played with men old enough to be my father and grandfather too. The older men were even keener at the game than the boys. Very nearly every day that I played I made a copper or two by selling back the marbles that I had won to the men or boys whom I had skinned.

Most of the miners' families had been for years dealing with the same grocer and butcher, whom they paid once a fortnight. During the whole of the strike the grocers and butchers allowed their old customers credit, trusting to their honesty to pay them back once the strike was settled and work in full swing again. Many of the customers did pay them back, but many didn't, so quite a number of grocers and butchers went bankrupt in South Wales that year or the next. It was very hard lines on families that did not have shops to fall back on, but most of them lived very well indeed, especially those that possessed kitchen gardens. They generally managed to raise the wind to buy sufficient bread for the household needs, and if they occasionally went a little short of butter and cheese they never went short of meat. There were thousands of sheep grazing on the hills around Blaina, and a good many hundreds of these found a happy resting-place in the stomachs of the strikers and their families.

At this time the South Wales miners were not in the Miners' Federation of Great Britain, though shortly after the strike was settled they joined it. The only money the strikers were getting was an occasional shilling from collections made for them in various parts of England. The parish granted no relief and even if it had done so it is very doubtful whether the strikers or their wives would have accepted it. People who took parish relief would have considered themselves disgraced for ever. South Wales was very proud and very independent and also very narrow-minded. Even the aged and infirm, and destitute widows, were treated with scorn if they accepted a little parish assistance, and I often heard it said, better to do away with oneself than go to the parish. But I must admit that things have greatly changed since that date. I think it is Socialist propaganda that is responsible for the change. People, rightly or wrongly, have come to regard parish relief or Government relief as their natural birthright and something to fall back upon if things go wrong. In 1921 (and on several occasions since) the heads of households who were drawing a high scale of relief came to be voted heroes of the first water. This strike at last came to an end and I went back to work with my old buttie, who had gained about a stone in weight. One dinner hour I overheard him telling the old soldier that he was so fed up with mutton that he had informed his wife that if ever she brought any of it into the house during the next twelve months she could look out for squalls.

There were only two political parties then, the Conservatives and Liberals: the Labour party was as yet in its infancy. The only general election I can remember as a boy was the one that took place just before I left school. At this election Sir William Harcourt, one of the great Liberal leaders, had been defeated in the seat he had contested elsewhere. West Monmouthshire was a stone-wall certainty for the Liberal party, so the Liberal member for West Monmouthshire in the previous Parliament, who was again contesting the seat, withdrew his nomination and Sir William Harcourt took his place. During his electioneering campaign Sir William visited Blaina and gave a speech at the Market Hall whch was applauded to the echo. Before he arrived at the Hall he rode in his carriage and pair for a good two miles through cheering lines of spectators. About a mile from the Hall some enthusiastic Liberals stopped the carriage and, taking out the horses, got between the shafts themselves and pulled their hero triumphantly along to the Hall. But my Uncle, who was a fervent Tory, and was wearing his party colours as conspicuously as possible, remarked as usual that the country was always a lot more prosperous under a Tory government than under a Liberal one. Sir William was returned by an overwhelming majority, which caused my Uncle to say that there were more damned fools in West Monmouthshire than even he ever suspected.

In October 1899 the Boer War broke out and David came home on a few days' leave before proceeding to South Africa. He was in magnificent condition and was looking forward to active service as a glorious adventure. My Aunt now felt different about David being in the Army and told her neighbours that she was proud that she had bred a son who had gone to Africa to fight for his Queen and Country. Before the Boer War it was commonly believed in the village that any young man who joined the Army did so either because he was too lazy to work or else because he had got a girl in the family way. Hardly anybody had a good word for a soldier and mothers taught their daughters to beware of them. The War had not been progress a week before everybody became wonderfully patriotic, and soldiers home on leave were feted like proper heroes. If a soldier walked down the main street of Blaina he was followed by admiring eyes and all who met him wanted to shake hands with him. Everybody sang "'Tis the Soldiers of the Queen, my lad" until they were blue in the nose, and another song, "Good-bye Dolly Gray," about soldier-boys a-marching and the bugles calling and hearts breaking, was murdered right and left.

When the news came through that Mafeking had been relieved I remember that my Aunt was so overjoyed that late in the evening she made a huge bonfire with four old straw mattresses on some waste ground in front of the house. It turned into quite a celebration. The flames attracted all the people in the street and some from the adjoining street as well; patriotic songs were sung and some of the women said that if they had been given notice that there was going to be a bonfire they would have made a Guy of Lloyd George to improve it. One patriotic old lady caused a good laugh when she said that she would like to improve it still more by using Lloyd George himself. Lloyd George had made himself very unpopular by his pronounced pro-Boer views, and he was lucky to get away with his life when he addressed a meeting at a large hall in Birmingham. The crowd was after his blood and he only escaped by the skin of his teeth, under the protection of a strong body of police and disguised as an under-sized policeman. They would have torn him to pieces if they had got their hands on him.

Politics are a farce – one has only to look at the way a man like Lloyd George goes up and down, or Ramsay MacDonald. He's the same man all along, namely a politician; but when I recall Lloyd George as the Guy on Mafeking night, and then again Lloyd George at the time of the so-called Khaki Election, in December 1918, when nothing in reason or out of it was too good for the Hero who had won the war and who would hang the Kaiser (if we gave him the necessary authority) as a sacrifice for the sins of the World – well, it makes me do a grin. And Ramsay

MacDonald, the "pro-German" and pacifist, whom patriotic people were desirous to bum as a Guy during the War, succeeded Lloyd George as the Saviour of his Country – this time not from the Germans but from the Reds. Personally, though I have lived ever since the War in a part of the country which has been severely hit by industrial depression, and have suffered severely from it myself, I can't help agreeing with my Uncle that a Tory government is likely to bring more prosperity with it than what a Liberal or Labour government ever brings. And in spite of all the Socialist propaganda that goes on about me I remain a rank Imperialist at heart: I am afraid I must have a kink somewhere.

I was growing into a tall, skinny lad and in 1900 I visited a recruiting-sergeant who lived a few miles away. I did my best to convince him that I was over eighteen years of age, but he would not believe me. He told me to clear off and pack twelve months' dinners in my belly, and then perhaps I would be big enough around the chest to pass the doctor. I waited another twelve months and meanwhile put on a lot of weight for a growing youth. I was now seventeen and a half years of age and big enough, I thought, to join any regiment in the Army. The war in South Africa was still in progress. I did not go to the recruiting-sergeant who had sent me away before but to Brecon Barracks, thirty miles off. The recruiting-sergeant there told me that I would pass the doctor in a canter; but he was disappointed when I told him that I wanted to join the Royal Welch Fusiliers. He did his best to persuade me that the South Wales Borderers were the finest regiment in the service – he wanted to know whether I had never heard tell of the Defence of Rorke's Drift. Well, then! – but all in vain, and he finally enlisted me for the Royal Welch Fusiliers. (My Aunt was disappointed too when she saw that I was not in the same regiment as David.) The reason I insisted on joining the Royal Welch Fusiliers was that they had one battalion in China, which had taken part in the suppression of the Boxer Rising, and the other battalion in South Africa, and a long list of battle honours on their Colours; and also that they were the only regiment in the Army privileged to wear the flash. The flash was a smart bunch of five black ribbons sewed in a fan shape on the back of the tunic collar: it was a relic of the days when soldiers wore their hair long, and tied up the end of the queue in a bag to prevent it greasing their tunics.

I passed the doctor with flying colours; the only comment he made was that I was a finely built lad but looked extremely youthful for my years. For instead of putting six months on to my age I had put on eighteen months: I was determined this time that if I was big enough I would also be old enough. The recruiting-sergeant never questioned my age and no birth certificate was required. If I had only been big enough the first recruiting-sergeant would not have questioned my age either: he would have been only too glad to pocket the fee that recruiting-sergeants received for each man they enlisted who passed the doctor.

I arrived at Wrexham, North Wales, which was the Depot of the Royal Welch, and my first night at the red-brick Barracks I spent in the receiving-room. The following morning I was examined by the Depot medical officer and on the 12th April 1901, I was a full-fledged soldier of His Majesty's forces – His, not Her, Majesty, because Queen Victoria had died that January, to the great and genuine grief of her people. I was given a rifle, a bayonet, and white buff straps with two white pouches which would hold about fifty rounds apiece. I then drew my soldier's kit from the Quartermaster's Stores, which consisted of one greyish-black top-coat, cape and leggings, one suit of scarlet, one suit of blue for drill and fatigue purposes, boots, shirts, socks, and other small necessary articles too numerous to mention. Every six months a man was entitled to a new pair of boots, and every twelve months to a new suit of scarlet, but he had to wait until he had completed twelve months' service before he was entitled to a proper scarlet tunic of more expensive cloth than the serge he got at first. He also was then issued with a busby, the tall fur head-dress reserved for Guards and Fusiliers. The ones issued about this time could have done with a little hair-restorer, but they were very light to wear. (An officer's busby, which was known as a "bearskin," cost some twenty or thirty pounds, I believe.) At the Depot we wore forage-caps[1] for all parades.

I was shown to a whitewashed barrack-room overlooking the square, where there were twelve other recruits and two old soldiers who had been abroad and were called "duty men". A corporal who had served sixteen years in the Regiment was in charge. A man called Toombs showed me the way to fix my straps for drill purposes and also told me what cleaning tackle I would require: I could buy it at the dry-canteen. He taught me how to fix my black kit-bag and top-coat on the rack over my bed and how the straps should hang on the pegs above the head of the bed. After he had given me a few further instructions and hints he said in an old-soldier's manner: "Well, youngster, when I first enlisted I had to watch what other men did and nobody gave me any hints or instructions about my straps or kit in anyway. From now on you want to keep your eyes skinned and you'll soon drop into the hang of things." Toombs had just completed three months' service.

During the evening Toombs showed me around the grounds of the Barracks, in which there were relics of former wars that the Regiment had taken part in. Outside the guard-room hung a large ornamental bronze bell which dated from the Burmese War of 1885, and in front of the Officers' Mess was a field-gun which had been captured during the Crimean War by Captain Bell, v.c., an officer of the Regiment. When we were outside the Canteen Toombs said he was sorry that he could not take me in to have a drink, as he was stony broke. Although young, I knew how beer

---

[1] In 1903 or 1904 the forage-cap was discarded in favour of the Broderick, as it was called, which was like a sailor's cap with a small piece of red in front, over which was worn the regimental badge. We thoroughly detested this new cap, which the majority of men said made them look like a lot of bloody German sailors; but we much preferred it to the poked cap which followed it, introduced, I think, by Lord Haldane. When this cap first appeared the men said that the War Office was being run from Potsdam. They were rotten caps to carry in a man's haversack.

tasted and was fond of a glass now and then. I told him to lead on and he could have a drink at my expense. I had about fifteen shillings and after a couple of pints he borrowed five of them, saying that he hated to see me paying all the time. It was my first appearance in a military wet-canteen, and everything seemed strange, including the weakness of the beer. Several men got up and sang songs, sentimental or comic ones, which were loudly applauded, but when one of the old soldiers out of my barrack-room rose to his feet to give one he was clapped and cheered to the echo. Toombs turned to me and said, "Now, youngster, you are going to hear a singer who in my opinion is second to none in the whole of the British Army. Before he has done singing I think you will agree with me." I did not expect to hear an opera-star but I did think that I was going to hear a singer out of the ordinary deliver a really classy song. I was sadly disappointed: the old soldier had a voice which resembled two rusty tin cans being slowly rubbed edgeways together. He sang a long ballad with a short chorus which was taken up with great gusto by the company present. I have heard some pretty far-fetched songs during my life but this was the king-pin of them all. It was called "The Girl I nearly Wed." I can only remember the tail-end of one of the later verses, which ran:

> *I wake up sweating every night to think what might have been,*
> *For in another corner, boys, she'd stored the Magazine,*
> *The Magazine, a barrel of snuff and one or two things more,*
> *And in another corner, boys, was the Regiment forming fours.*

The chorus, four times repeated, was: "She was up the bleeding spout." When he finished, the singing was so rapturously applauded that he was forced to give an encore, which he did with "The Ballad of Abraham Brown and the Fair Young Maiden," which was something to bring a blush to the cheek of the most hardened prostitute that ever lived.

In the midst of this song, not all the words of which I

understood, two men raised their voices in argument, but immediately became quiet when the temporary chairman roared at them: "Order there, you lousy bastards, or I'll come over and knock hell out of the both of you. Have you no respect for music?" The chairman, who was a duty man, was handy with his dukes, Toombs said, and wouldn't allow any nonsense.

The Canteen opened at noon for three-quarters of an hour, and again from six to a quarter-past nine. At stop-tap Toombs and I were quite sober, which made me remark that one could drink a great many glasses of this sort of beer without feeling the effects of it. "You are quite right," he replied, "that old bastard in charge of the Canteen has been using the water-can on it to some order today. But if I'm not on duty tomorrow evening I'll take you around the town, where a man can get a decent drink of beer, though it costs double what it does in Barracks." (Beer was only three-halfpence a pint in the Canteen.) We walked out in the town the following evening, each with his regimental swagger-cane without which we were not allowed through the gates, and coming along the road Toombs said that we would visit a pub where a few respectable clean skirts frequently called. He said that he was quids in with one of them and that she generally had a companion with her whom he could introduce to me. Up to then I didn't know what women were – that is, to have sexual intercourse with them – but my companion, who was four years older than I was, knew all about them, or so he thought, and told me during the course of the evening that he enjoyed nothing better in the world than a nice bit of skirt. I was so innocent that I honestly did not know what he meant by "respectable clean skirts," but I did not give my ignorance away: I pretended to be as knowing as he was.

We had been in the pub about an hour, sitting in one of the side-rooms, when in walked two young women, who greeted Toombs in a friendly way and asked to be introduced to me. After a drink or two Toombs was soon in deep private conversation with the one of them who, he had told me when we were out at the back together, was his own bit of goods. She was a married woman whose husband was away working and only came home every other week. I was soon paired off with the other one whose husband, so she said, was also absent from home. Her conversation pleased me very much and when, just before dusk, she said that she would be glad to go out for a breath of air, I volunteered to accompany her. The room was certainly getting stuffy. She whispered to her companion and a few minutes later, by arrangement, they left the pub together. We followed about twenty yards behind until we were on the outskirts of the town, where we paired off again. We walked a little further and my lady-friend then suggested a rest. We went off the road into a field, and my companion with his lady went on another fifty yards before settling down in the same field. The respectable clean skirt soon found out how innocent I was: we had not been resting more than a minute before she began to kiss and caress me. I laughed and told her to stop it, and made a half-hearted attempt to get away from her which only made her kiss and caress me more. In less than no time I had lost my senses and also my innocence. I have been with many women since then, some cold and some warm, but though it may have only been that she was the first, I fully believe this one was the warmest of them all. About an hour later I was brought back to my senses by Toombs shouting: "Come on, youngster. Just time enough to get back to Barracks." I parted from my lady-love, promising to meet her on the following evening, when she said that she would not have her companion with her. As we hurried back to Barracks Toombs said I was a lucky devil to drop on such a lovely clean skirt, and that he and several other men had tried to walk her off but there was nothing doing.

Toombs was a good friend to me. He taught me to recognize the various bugle-calls that were continually sounding from Reveille to Last Post, and also warned me of the usual tricks that are played on recruits. For instance, he told me that if ever I made my bed in the evening before going out for a walk in the town I was to examine it carefully on my return in case it had been "set." The bed was an iron cot, supplied with biscuits (straw-filled mattresses), blankets and a couple of rough sheets which were changed once a month. The cot was in two halves, one of which was made to slide under the other, to allow more space in the room by day. The two halves could be disconnected and then balanced together in such a way that they held up until the owner tried to lie down: then they collapsed with a crash and his backside got a nasty jar. My bed was set for me on my third night and I spotted it at once. I then played a trick on the trickers by once or twice pretending to be just about to sit down on it, and then remembering something and putting off my intention. Finally, when I was undressed, without saying a word I lifted off the bedclothes, connected the two halves again securely, put the bedclothes back, got in between them, and was soon fast asleep.

One of my fellow-recruits when he had only had two days of service was called up by an old duty-man and asked whether he knew where the Guard-room was. When the recruit proudly said that he did, the old soldier asked him, for a favour, to go to the Sergeant of the Guard and beg him for a lend of the whitewash brush. Off went the recruit, but presently he returned and said that the Sergeant wanted to know what the brush was needed for. "To whitewash the Last Post, of course," the duty-man replied with a solemn face. The Sergeant of the Guard then informed the recruit that he would require a special brush for that purpose; maybe he could get it at the Quartermaster's Stores. But the store-man there sent him on to the dry-canteen, and the dry-canteen sent him on to the Canteen, and from there he was sent to the Married Quarters, and then on to the Sergeants' Mess; and that was as far as the joke carried him, because all the Sergeants burst out laughing in his face. This was the oldest joke in the Army, but there were always new recruits to play it on.

Toombs soon left with a draft for Plymouth. He told me how sorry he was that he could not pay me back the five shillings he had borrowed from me. I told him to forget it, but months after, when I met him at Plymouth, he insisted on paying me in full, saying that he had just had a good win at cards. Toombs passed out of my life when he went to South Africa and I have never seen him from that day to this. During the rest of my time at the Depot I met my lady-friend pretty often and we had some delightful evenings together. I was sorry when I had to say good-bye to her; but I soon forgot her.

The Depot staff consisted of a colonel, an adjutant, a few young officers, a sergeant-major, drill-sergeants and corporals and about a dozen old soldiers. One of the old soldiers, called Pond, had over twenty-one years' service. It was said that during the first twelve months of his soldiering he and another man had deserted from Dum Dum in India where the First Battalion were stationed at the time. They had made their way for over a thousand miles towards the North-West Frontier, when Pond's pal died. Pond buried him and then made his way back to the nearest military station and gave himself up. He was tried for desertion and served a term of imprisonment. After he came out of prison he turned over a new leaf and proved himself a model soldier. After he had completed his eight years with the Colours he was allowed to make it twelve years and later to extend his service to twenty-one years. He had lately completed his twenty-one years but had been allowed to extend his service for another four years. In 1914 when I rejoined the Regiment as a reservist Pond was still pottering about the Depot. By that time he had completed thirty-five years' service, and the true facts of his early adventure had been forgotten: it was now believed that he had deserted on the North-West Frontier and gone with his pal through Afghanistan, eventually surrendering to the British Consul at Jerusalem. He was a moody old chap who would never speak of his experiences in India or anywhere else, and it was seldom that he opened his mouth to speak to a recruit at all.

In summer Reveille was blown at half-past five, in winter at half-past six. As soon as the bugle sounded from the Guardhouse the Corporal in charge of our room would shout: "Show a leg, get out of it, open those blasted windows and down with that p—s-tub!" If any man rolled over a few times before getting up, the Corporal would seize hold of his bed and pitch him out on the floor. Each recruit took his turn as orderly man for the day. His duties were to draw the rations, bring the food from the cookhouse and wash up the plates and basins after meals; he had also to take the urinal tub away in the morning and bring it up again in the evening, the tub being placed on the landing just outside the door. The tub was always very full on the morning following pay-day and that day's orderly man had to have assistance to carry it down. A man's daily rations consisted of one pound of bread, three-quarters of a pound of meat, vegetables, tea and sugar. We lived, dined and slept in the same barrack-room, which we had to scrub out twice a week with hot water and soap.

Our first parade was before breakfast, for forty-five minutes. The last quarter of an hour consisted of violent exercise, such as running with knees up like the high-stepping carriage-horse of that period. That gave us such an appetite that most of us used to consume our entire pound of bread for breakfast, at the risk of feeling equally hungry at tea. Immediately after breakfast a few men were told off to scrub the tables, while others dry-scrubbed and cleaned the room before going on the second parade. The pay was a shilling a day, from which was deducted threepence a day for messing and a halfpenny a day for washing. Out of this threepence a day was found for extras, which on week-days consisted of about an ounce of butter a man, to do for his breakfast and tea; on Sundays we had liver and bacon for breakfast, or two boiled eggs. A corporal was generally in charge of the mess-book and the extras were bought from the dry-canteen. Every corporal that I can remember managed to get the mess-book in debt at some time or other, which meant dry bread for breakfast and tea until it was back in credit again.

The washing was done by the wives of the corporals and old soldiers who were married on the strength of the Regiment and living in barracks. We were only allowed to send shirt, towel and socks to the wash, which the married men collected once a week. Men and wives married on the strength of the Regiment were called the "married crocks." One had to have five years' service and be twenty-six years of age before one could get married on the strength. A regular number of married men were allowed in each regiment and an application to get married on the strength had to be presented to the Colonel. If the regiment had its required number of married couples it was a case of patiently waiting until a vacancy occurred. The married men on the strength were generally men who intended to do their twenty-one years, or more if they could. They were allowed free quarters, coal and light, and their wives and children also drew daily rations. A man who married off the strength had to keep his wife out of his own shilling a day; she lived outside barracks and he inside, and they met whenever they could, but officially she did not exist. I can only remember two men who got married off the strength; both went to India, where they had to serve seven years before they returned to England. Seven years is a long time for a young married couple to be separated; much could happen in that time.

We were paid five shillings a week and settled accounts with the Colour-Sergeant on the last day of the month. I can't remember during the whole of my soldiering at home ever stepping forward to sign my accounts at the end of the month without finding a considerable deduction made for barrack-room damages. The monthly deductions sometimes ran to a shilling a man: they were supposed to cover the loss of barrack-room utensils, plates and basins, and the supply of soap and floor-cloths for scrubbing the rooms. Barrack damages were always a mystery to soldiers serving at this time. I once ventured to ask the Colour-Sergeant at the pay-table for an explanation of them. He obliged me with a list of things which had been supplied my own barrack-room, enough to stock one of Woolworth's stores if they had been introduced into England at this time; and yet the only utensils that had come to our notice during the month were a fourpenny black-lead brush and a bar of yellow soap. I knew it was useless to

19

make a complaint to the paying-out officer or company officer; the Colour-Sergeant would have explained that though it might be so that my room had incurred few damages that month, the other company rooms had been exceptionally unlucky, and that, as usual, the money was being made up by the whole company contributing the same share. I always believed, but could never prove it, that the colour-sergeants and the men in charge of the dry-canteen, where the barrack-room articles were bought, had a perfect understanding. No doubt they settled their accounts once a month and drank to the health of Barrack-room Damages.

After breakfast, we recruits did two drill-parades of forty-five minutes each, with a fifteen-minute rest after each one. The third parade of the morning, which lasted an hour, was in the gymnasium; and by the time the physical-drill instructors had finished with us, the majority of us were hungry enough to eat a brass monkey, tail and all. If we had any money for a pint or two in the Canteen, the beer would make us hungrier still, and many a day I felt that I could have eaten the whole of the men's dinners in my room. At a quarter to one the dinner was brought from the cookhouse. It was either a roast, an Irish stew or a curry stew, with occasionally a small portion of currant-pudding as a dessert. It was the orderly man's job to make it out on each man's plate as equally as he could. One or two men assisted him, with the remainder of the men in the room standing around the table and looking on with roving eyes and hungry bellies. After the plates were all filled and the dixie emptied, the orderly man would say, "If you are all satisfied, charge!" Although all the dinners had been made out as equal as possible, there would be a charge at one of the plates which seemed to have a fraction more on it than the remainder. The rule was that the man who first laid his hands on it kept it, but it sometimes happened that there was a dead-heat. This generally produced a fight and the victor took the spoils. The Corporal and the two old soldiers were rarely in the room when the dinner was made out; if they were, they never interfered with our charge and were delighted whenever a scrap took place. I had one in my first week over a bit of roast which another recruit took a fancy to; but boxing practice at Blaina with Evan and David (David had now become middle-weight champion of his battalion in South Africa) had put me well above the average in skill. I knocked my opponent out almost at once and from then on I was left severely alone; indeed I never again had the necessity of scrapping with a man of my own regiment, and did not go about picking quarrels.

We did two more parades in the afternoon: on one of these parades we were occasionally lectured on the history of the Regiment and had all the battles which were emblazoned on the Colours fought over for us – and some that were not, such as the Battle of the Boyne which, taking place in a Civil War, did not count as a battle-honour, and Bunker's Hill, where the Regiment fought gallantly, but lost four-fifths of its strength charging in the heat in full equipment up a difficult hill against American sharp-shooters. The battle-honours we were taught to be most proud of were: Minden, where the Regiment was in the front-line when six British infantry battalions drove twice their number of French cavalry off the field; Dettingen, where King George II fought in person; Albuhera, where our men stormed the heights, as part of the famous Fusilier Brigade; Corunna, where our Second Battalion were the last troops to embark after the victory; Waterloo, where the commanding officer moved the square up out of the reserves on his own initiative and was himself killed; and Inkerman, the "Soldiers' Battle" as it was called, in the Crimean War.

Every Friday afternoon we had to do one hour's saluting drill, which every one of us detested, from the time we began to learn it by numbers. N.C.O.s, representing officers, were stationed at different points around the square, around which we marched four abreast with about ten or fifteen paces' distance between each four. When we were so many paces from an N.C.O., the left-hand or right-hand man of the four would shout "Up!" and each man would bring his hand smartly to the salute, and keep it there for so many paces until the command "Down!" when the hands were brought smartly to the side. Any man who was judged to have been slovenly in his saluting was punished with an extra hour's saluting-drill after the parade had been dismissed. I never did this extra hour myself and always considered it the most childish and degrading punishment of all. I have often seen one solitary man going through it under charge of an N.C.O. Since the other N.C.O.s had been dismissed, the man had to walk around the square saluting the trees which enclosed it.

At night, we had to stand by our beds at nine-thirty when the Orderly Sergeant came around the rooms and took down the names of absentees. An hour later the bugler sounded "Lights out" and every light was extinguished. After a man had been in the service six months without an entry in his regimental conduct-sheet – the company conduct-sheet was only for light offences – he was entitled to a permanent pass which allowed him to stay out of barracks until midnight, when he would answer his name to the Sergeant of the Guard. If a man was late, a lot depended on whether he was on friendly terms with the Sergeant of the Guard, who might let him get away with it even if he came in at two o'clock in the morning, or who might, on the other hand, report him as "absent until reporting himself at 12.5 a.m." if he was just beaten at the gate by the last stroke of midnight. Men did not have enough of money to square the Sergeant of the Guard, but I daresay it could have been managed if they had.

My own length of service did not entitle me to a late pass, but Toombs put me up to the trick of getting over the brick wall at the back of the Barracks, with the assistance of a handy tree, and spending the night with my fair charmer. I never failed to be back in my bed by reveille, to start another day as a young soldier should, running around the parade ground with my knees up to get an appetite for my breakfast.

RECRUIT LIFE IN 1900

At the end of three months I left Wrexham with a draft for Crownhill Barracks, Plymouth, where our regimental details were. I hoped that the time was not far distant when I would be under orders to proceed to South Africa, where my cousin David was still going strong. He seemed to enjoy the War and had been transferred to the Mounted Infantry. I had learned much during these three months, and as I had done several guards and pickets and was also considered an expert at rolling a top-coat I felt myself a pukka old soldier. There was quite an art in rolling a top-coat, which was strapped on the small of the back when parading in full marching order. Two or three men assisted in the actual rolling, but the art lay in arranging the coat on the floor or table beforehand.

Although the War was in its last stages, and casualties from enemy action were no longer serious, drafts were still being sent to our First Battalion, to make up for the wastage caused by sickness. A few First Battalion men also completed their engagements and were able to go home, but not many; because while the War was in progress time-serving soldiers, not only in South Africa but everywhere else, were kept with the Colours until they had completed their full reserve service, which meant thirteen years in all; or twenty-one years if they had re-engaged for that period; or seventeen years in the case of Section D men. No drafts, however, had gone to our Second Battalion in the East for some time. I may mention here that the Indian Government realized that as soon as the War ended large numbers of seasoned troops would be demobilized as time-expired, and untrained ones sent out to take their places; so they insured against too sudden a change by offering a bounty of £ 26 10s to all men who had completed eight years' service if they would take on for their twelve – twelve really meant thirteen, as in bakers' shops. (They offered at the same time a three-months' furlough to England, but that reduced a man's bounty by £ 12 or so, so few men accepted it.) This wise move on the part of the Indian Government explains why so many old soldiers were still serving with the Second Battalion long after I joined it in India when the War was over.

Crownhill Barracks was about three miles from Plymouth, which I found a very lively place. The principal street, called Union Street, was lined with pubs which in the evening were full of sailors, soldiers, marines and prostitutes. Some of the pubs were never called by their proper names: there was one in a little side-street just off the lower end of Union Street that was known as "The Whores' Canteen" and another had a name which even the prostitutes resented. Some of these women were young and good-looking and many of them had nicknames, but according to soldiers who had been stationed at Plymouth two years before there was not one of them now with one-quarter of the good looks that a girl called Klondyke Nell had; she had vanished mysteriously from Plymouth, and Union Street hadn't been the same place since. They were always talking about this Klondyke Nell, but it was no use, she was never seen or heard of again. From all I heard of her she was in the same class as that Eskimo Nell who defeated Deadwood Dick and Mexican Pete in the famous ballad which bears her name; she certainly left her mark on Plymouth in the year 1898.

I was in many a scrap in these pubs. Affrays between men of the Regiment were fought out more or less according to the rules of boxing, but when we were scrapping with marines or men of other regiments with whom there was a feud that was a different matter: the buckled ends of belts were used, and also boots. Soldiers and marines were always at loggerheads. The Welsh Regiment had a particular feud with the marines and if a row

24

started and the Welsh were in a minority we of the Royal Welch Fusiliers felt bound in honour to go to their support. The first I knew of the business was, one day in a pub, a man of the Welsh Regiment went over to a marine and said in a friendly sort of tone: "Pleased to meet you, Joey, let's you and I together have a talk about old times."

"What old times, Taffy?" asked the marine, suspiciously.

"That sea-battle long ago – I forget its name – where my regiment once served on board a bloody flagship of the Royal Navy."

"What as? Ballast?" asked the marine, finishing his beer before the trouble started.

"No, as marines, whatever," answered the Welshman. "It was like this. The Admiral wanted a bit of fighting done, and the sailors were all busy with steering the bloody ship and looping up the bloody sails, see? And the marines said they didn't feel like doing any bloody fighting that day, see? So of course he called in the Old Sixty-Ninth to undertake the job."

"Never heard tell before of a marine who didn't feel like fighting," said the marine, setting down his empty mug and jumping forward like a boxing kangaroo. In a moment we were all at it, hammer and tongs, and the sides being very even a decent bit of blood flowed: fortunately the scrap ended before murder was done, by the landlord shouting that the picket was on the way. We made ourselves scarce, friends and foes alike, sneaking down an alley at the back of the premises.

The troops from the barracks found main-guard and pickets for six days of the week; the marines found them on the remaining day, which was Saturday. All drunk and disorderly prisoners were put in the main guard-room, where an escort came for them the next morning. The marines were bolder on Saturday nights than what they usually were, because there was not the same danger of being rushed away to the guardroom if they were found scrapping, as would certainly happen, whether they were drunk or not, if their opponents were of the same regiment as the picket.

On Saturday nights also, soldiers were particularly careful when passing a picket of marines: there might be old enemies among them. The marine pickets always made the most of their one night in seven for paying off old scores.

The same sort of trouble occurred if Scottish and Welsh troops were in the same garrison together. We and the Highland Light Infantry were bitter enemies, I don't know why – it was something handed down from bygone days. Some say that it originated towards the end of last century during a final for the Army Football Championship of India when the H.L.I. having scored a lucky goal early on against our chaps, kept their advantage by delaying tactics – kicking wide into touch whenever they had the ball. To this day, in the Battalion, these tactics are always greeted with the indignant cry of "H.L.I., H.L.I.!" and the expression has been adopted by other units and by civilians. We got on well enough with the Irish regiments, but the only Scottish regiment that I can remember us ever being friendly with was the Cameronians.

A good deal of rhyming-slang was used in those days. For example, a pub was a "rub-a-dub," a table was a "Cain-and-Abel," the wife was "joy-of-my-life," the kids were "Godforbids" and so on. Beer was "pig's-ear" or "Crimea" or "Fusilier," but if a Welshman went into a pub where a Highland soldier was, of the regiment whose square was once broken by the Mahdi's dervishes in the Sudan, he would sometimes ask for a "pint of broken-square." Then he would have his bellyful of scrapping for the rest of the night, because this was an insult that the Highlanders could not forgive; they swarmed up at it like ants. If a man wanted a scrap and couldn't think of a suitable insult, all he had to do was to turn his beer-mug upside-down on the bar as soon as it was empty, which meant a general challenge.

With the sailors of the Royal Navy we always got on very well. There was no sense of competition between them and us, as there was between us and the Marines or men of other regiments. What struck us most forcibly about the sailors was the comradeship between petty-officers and ordinary able-bodied seamen; they walked out and mucked-in together when taking an evening ashore just as if there was no difference of rank between them. Such behaviour was sternly prohibited in the Army. From the day that a man was made a lance-corporal he could no longer walk out or gamble or drink with a private, or allow himself to be addressed by a nickname except by his equals. If he forgot the responsibility that his stripe carried with it he was immediately crimed for "conduct unbecoming a non-commissioned officer" and would probably get a severe reprimand from the Colonel, and be reduced to the ranks again. It was not wanting to lose touch with any chums that always kept me from putting in for promotion (though a few of them lost touch with me by doing so) and I remained a private to the last, in spite of all temptations.

It was at Christmas that most fighting was done at Plymouth, the same there as in every station I was ever in. Christmas always meant a damned good tuck-in, with plenty of booze and scraps to follow. On this day stern N.C.O.s winked their eye at everything: a man had to be raving mad before they would rush him along to clink. Men who had the money would get dixies full of beer from the Canteen; even the Provost-Sergeant winked his eye at it. During the afternoon, when the beer took effect, scrapping started and unpopular N.C.O.s made themselves scarce for the remainder of the day. It was the custom to take the battalion out for a fifteen or twenty mile route-march, two days later, to sweat Christmas out of them.

From Crownhill we made a move to Raglan Barracks, Devonport, not far away, where we formed a composite Battalion with details of the Welsh Regiment, Gloucesters, Duke of Cornwall's Light Infantry and Somersetshire Light Infantry. This battalion was commanded by Lord Mostyn, a Colonel of one of our Militia Battalions; the Adjutant and Regimental Sergeant-Major belonged to the Duke of Cornwall's Light Infantry. We carried out the usual drills together and went on a fourteen mile route-march once or twice a week. Each unit in its turn found quarter-guard, hospital-guard and mainguard. The main-guard was on the main road, not very far from the halfpenny bridge which divided Plymouth from Devonport. On this guard one sentry was posted over the house of Lieutenant-General Butler, who commanded the Western District, and another over the house of the Lord High Admiral, or whatever his official title was. This guard had to learn the distinguishing marks on the Naval Officers' uniforms and give them the salute they were entitled to.

The first time I was posted as a sentry over the Admiral's house, I noticed two small boys on a flat roof at the end of it; they had clay pipes and a bowl of soap-suds and were trying to beat each other at making the largest bubble. I became so interested in the game that I hardly noticed the approach of two civilians until they were a few yards away from me. They were both typical Devonshire farmers, the elder having mutton-chop whiskers and the younger being a burly, slow-looking man. They seemed to be even more interested in the boys than I was, but the boys were so excited in the competition that they never once looked down. The three of us stood watching for a few minutes together until I threw my rifle to the slope and prepared to march to the end of my beat and back. As I left them I told them that I wouldn't have thought it possible that soap-bubbles could provide so much excitement and that it would have been possible for the Lord High Admiral and his Staff to have passed my beat without me noticing them. They both burst out laughing, but even their laughter did not attract the attention of the boys who went on blowing bigger bubbles than ever. I marched to the end of my beat and as I was retracing my steps I was surprised to see the two farmers half-way through the front door of the house. About ten minutes later one of the servant girls, with whom I became very friendly afterwards, came out. She told me that the farmers were the Admiral and his son-in-law, a Commander in the Navy, the grandfather and father of the boys, and that they were both still laughing at what I had said.

The principal Medical Officer of the whole of the Western

District was a Surgeon-General nicknamed "Mad Jack." Men who had been admitted to the station hospital said that he was not a bad sort, but they all swore he was stone mad. The chief trouble seemed to be his familiar way of talking to the men. There were a number of sick and wounded men from South Africa in the hospital and whenever he visited the wards they were in he would smile and say, "My poor South African heroes!"; but when he entered the venereal ward he would smile and say, "My poor Union Street heroes!" In each case he would ask them how their wounds were healing. Hospital guard was popular: we were allowed a liberal supply of bread and cheese and hot cocoa for supper by the hospital authorities, and most of us were hungry-gutted youngsters.

I fired my recruits' course of musketry at Tregantle Fort, about six miles from Plymouth and not far from Eddystone Lighthouse. I proved to be a good shot and three months later when I fired my duty-man's course with the company I became a marksman. With the exception of one year, when I missed it by three points, I kept my marksman's badge through-out the whole of my soldiering. A duty-man's course of musketry had to be fired once a year. At this time twenty-one rounds were fired at two hundred yards (seven standing, seven kneeling and seven, rapid, lying), fourteen rounds at three hundred yards (seven kneeling and seven lying) and seven rounds each (lying) at five hundred, six hundred and eight hundred yards. At two hundred yards we also fired seven rounds (standing) at the head and shoulders of a man on a target which was exposed for three seconds and seven rounds at "the running man." This target was mounted on a trolley and pulled across the butts at a fair rate until it disappeared out of sight. Money-prizes were awarded, ranging from a sovereign to half-a-crown, but were abolished a few years later. A marksman generally won a money-prize and was also entitled to wear the cross-rifles on his sleeve. A first-class shot came next to a marksman, followed by a second-class shot; lowest of all was a third-class shot. At the Tregantle range firing would often be held up for half an hour or more owing to a fishing fleet sailing too close to the shore; this was to protect the boats from casualties which the third-class shots were likely to inflict.

During the night it was easy to smuggle anyone into Raglan Barracks but they had to be out before dawn, because by day escape was impossible. The barracks were three storeys high and surrounded on one side by a very high wall; on the other, facing a common called the Brickfields, ran a stone wall about seven or eight feet in height with iron railings sunk in the top. Sometimes girls were brought into the barracks at night over these railings and sent out over them an hour before dawn. The room I was in, on the top storey, was only partly occupied at this time – six of us and the Corporal in charge. The Corporal had a bunk of his own partitioned off at the end of the room. At about half-past eleven one night the Corporal and another man from the room brought in a young lady whom they had picked up in town. She had only just begun in her new profession and was half-drunk. The Corporal, who was an old soldier and a happy-go-lucky chap, took her into his bunk and slept with her. He intended to get her out of barracks before dawn, but in case he did not wake up himself anyone of us who happened to be awake at the time would give him a call.

Reveille was sounding before the Corporal or anyone else of us woke up; it was now impossible to get the lady out of barracks. The Corporal told us not to breathe a word to the men in the other rooms about the matter; he would get her out as soon as dusk set in. Once a day the Company Officer inspected the barrack-rooms, but it was very rarely that he entered the bunks of the N.C.O.s in charge. The Colour-Sergeant was a very stout man and only paid a visit to the upper storey when he came around with the Company Officer. In case the Officer did go into the bunk the girl would be hiding under the bed; for N.C.O.s in charge of rooms did not fold up their beds in the morning like the men did. Our only difficulty was what to do about sanitary arrangements. The sanitation was situated on the ground floor

30

and the young lady wanted to know what she should do if she had a call of nature.

One man suggested bringing up a tub but this suggestion was turned down; if one had been brought up it would have been noticed by the men in the other rooms who would have smelt a rat and perhaps investigated. Another man had a bright idea: the spare tea-bucket! The orderly man could take it down that night as soon as she was gone, and clean it. We all thought this was a wonderful solution, with the exception of the orderly man who thought it was a rotten one; but we appealed to his sporting instincts, and he finally promised to do the job. Each room had two tea-buckets, one in use and the other in reserve. Our pair, unluckily as it turned out, happened both to be brand new. It was the orderly man's duty to take the tea-buckets down to the cookhouse before breakfast and again before tea. The cook put the numbers in whitening on each room's bucket so that each room would receive its proper one.

The Company Officer inspected the room in due course, with the Colour-Sergeant, and everything passed off all right. The Corporal did not mind anyone of us in the room visiting the girl during the day and neither did the girl. In the course of the morning the Corporal was warned that he would be on picket in the town that night, which meant that he could not be there to assist her over the railings. He told us to do so for him as soon as darkness set in. The evening came, and we told her to get ready; we gave her an army top-coat to cover her dress and a forage-cap to put on her head in case we met anyone going down the stairs or crossing the square. She had come into barracks this way too. But she threw the coat and cap back at us, saying that she enjoyed barrack-life with so many nice young men about and that she was not leaving that night, and that if we got rough with her she would scream the barracks down.

About midnight the Corporal returned and we explained the situation to him. He also tried to persuade her to go, but in vain. In the end she promised to leave the following night. We again successfully appealed to the orderly man to carry on the good work. The Corporal now had the wind up; if the girl had been found in the room he would have been court-martialled and reduced to the ranks, with a stiff term of imprisonment thrown in. She remained there all the following day, hiding under the bed during the Company Officer's inspection, and receiving visitors. When night came, however, she kept her promise and we got her safely over the railings. It would have been very awkward the following day if she had not gone, as it was the day for the inspection of barrack-room utensils: every utensil in the room, including both tea-buckets, had to be laid out for the Officer's inspection, and it was the orderly man's duty that day to polish the tea-buckets bright enough for a man to shave in. During the inspection of our room the Officer complimented the Corporal on having the two brightest tea-buckets he had seen that morning and said that they were a credit to the man who had cleaned them.

In the afternoon, when it was time to take the bucket down to the cookhouse for the tea, the orderly man found that he had cleaned both of the buckets so well that he was unable to decide which of them had been used for another purpose. No bloodhounds that ever followed a trail could have ever sniffed harder than what we seven then did; the buckets were passed and repassed around and we sniffed every inch of the inside and outside of them. One man was so hot on the scent that he got his head fast in one of the buckets through trying to get his nose as near to the bottom of it as he could; when we released him he said he was not sure, but the thought that was the bucket in question. We gave the problem up as hopeless; there wasn't a vestige of smell in either of the buckets. We cursed the orderly man for doing his work so well and for not thinking of putting a small nick in the bucket; and he cursed us back for not having thought of the nick ourselves. He took one of the buckets down to the cookhouse and brought it back to the room filled with tea. Then he poured out each man's portion in his basin. We all again gave an exhibition of sniffing that would have taken some beating, but in the end the tea went down untasted.

The Corporal was very pally with one of the cooks, who was also an old soldier, and revealed to him the secret of the buckets. The cook did a grin and promised to see to the matter. The following morning one of our buckets was taken down to the cookhouse, and another sent back in its place, not quite so new, but very welcome. We drank our tea this time without sniffing and during the afternoon the same thing happened with our other bucket. We had a great weight off our mind, and what happened to the rooms which now had our tea-buckets we did not care; they were probably congratulating themselves on having brand new tea-buckets instead of their old ones; for "what the eye doesn't see, the heart doesn't grieve," and also "no name, no pack-drill."

During the latter end of 1901 or early in 1902 King Edward and Queen Alexandra paid a visit to Plymouth and Devonport. They were present at the launching of a new battleship at Devonport Dockyard which Her Majesty christened The Queen. There was a grand procession down Union Street, with troops lining both sides of the road to keep the crowds back. Their Majesties had a wonderful reception, with cheers rolling from one end of the route to the other. There were none in the crowd who cheered more sincerely than the prostitutes did. They had turned out in force to witness the procession, forming pickets all along Union Street. Their Majesties were riding in an open landau and the Queen looked very youthful for one of her age, and very lovely also. King Edward was a jovial-faced man, very popular with the rank and file of the Army. He knew the history of every regiment and was said to be the greatest living authority on details of military and naval uniform. He had already won two Derbys and was also supposed to be very fond of a gamble at a game of cards, like the majority of his loyal troops. The Royal carriage had just passed the spot where I was standing when a member of one of the prostitute-pickets wriggled through the troops in front of her and shouting "My Father, my Father" started running after the carriage, which she had just failed to catch. This girl, who was called Leaky Minnie, was not quite right in her head. She was always telling her friends that her father was a Lord, but when she got drunk a Lord was not important enough for her: she then told them that it was either a Duke or a Prince that was her father. On this occasion the sight of the King must have played on her mind already stirred up by gin and patriotic feelings. The troops were taken by surprise, but a couple of men grounded arms and rushed after her. They caught her before she could catch hold of the back of the carriage and handed her over to the civilian police. She was never seen in Plymouth again and although her friends made enquiries they could never find out what had become of her. There was no mention of this incident in any newspaper the next day.

This winter we did a number of route-marches in full marching order through the villages surrounding Plymouth. Any man who fell out was punished the following afternoon by having to march for two hours in full marching order around the large barrack square. In the middle of January 1902 about twenty of us chosen from the different units of the battalion commenced a sixty days' course of signalling under a signalling sergeant of the Gloucesters. He was assisted by an old signaler of our Second Battalion who had been invalided home from China. As I found out later a signalling-sergeant always considered himself lucky if at the end of a sixty days' course he could pick out five men from a class of twenty who were likely to make decent signallers. This class proved no exception to the rule. I was greatly interested in this course, but the majority of the class had put their names down for it only because for the time being they would be getting out of drill-parades and guards and pickets.

It was while I was on this course that I committed my first crime. "Crime" in the British Army is a word that covers all offences from desertion or murder to being two minutes late for parade or failing to polish up one's cap or collar-badges to the

34

satisfaction of the Company Officer. My crime was as follows. As I was now not doing drill-parades I had poured some rifle-oil down the barrel of my rifle, which was the correct thing to do when it was not in use. My company commander was a captain belonging to the Royal Engineers Militia, and in the course of one of his weekly tours of the barrack-rooms he happened to pick hold of my rifle and squint down the barrel. The oil had dulled the grooves of the rifle and the Captain, who did not know the difference between a dull rifle bore and a dirty one, exclaimed: "Colour-Sergeant, this rifle is in a filthy condition. Have a look down the barrel!" The Colour-Seregant obediently squinted down the barrel and agreed with him, because he was just that sort of Colour-Sergeant. If the rifle had been in a deplorable condition of rust and dirt and the Captain had said it was remarkably clean, the Colour-Sergeant would have agreed with him exactly the same. I was on parade at the time, and when I came off the Colour-Sergeant informed me that I was a prisoner. This expression, which was officially abandoned during the Great War in favour of "accused" merely meant that I was charged with some offence: it did not put any restrictions on my movements. He warned me to appear at the Company Orderly Room the next morning to answer the charge of having a dirty rifle.

The following morning I appeared in front of the Company Commander. He told me I should be ashamed of myself for leaving my rifle in such a dirty condition in the barrack-room and then asked me what I had to say for myself. I said that the rifle was dull, not dirty, and that I had only been following Musketry Regulations for the proper care of arms when not in use. He did a sarcastic grin at my explanation and said that he would punish me lightly by fining me one shilling, which it would cost to have the rifle cleaned by the Armourer-Sergeant. This pretended kindness made me angrier than what I would have been if he had said nothing but "Seven days confined to barracks." I told him I would not accept his punishment: I was not going to pay a shilling to have a rifle that was not dirty cleaned by a special process. He was surprised by my obstinacy and put me back to be tried by the Colonel. Officers commanding companies could give no higher punishment than seven days confined to barracks. Punishment higher than this was awarded by the Colonel, who could sentence a man from eight days confined to barracks up to twenty-eight days' cells. If still more severe punishment was called for it was awarded by a District Court-martial. If a man was not satisfied with his company commander's punishment he had the option of being tried by the Colonel; if he was not satisfied with the Colonel's award he could claim to be tried by a District Court-martial. It was very rarely that this option was claimed; a colonel always awarded a harder sentence than a company commander, and a District Court-martial sentence was harder than a colonel's.

If any civilians had been present when I was marched into the Orderly Room in front of the Colonel they would have taken me for the most dangerous desperado in the whole of the British Army. The quarter-guard always provided an escort for prisoners and I was marched into the Orderly Room between two men who carried unsheathed bayonets by their sides with the points sticking upwards. The Captain gave his evidence and by the time he had finished it would have taken the Armourer-Sergeant at least a week to restore my rifle to working condition. I then replied to the charge. I was foolish enough to suggest that in fairness to me the Colonel should examine the rifle himself. I had by this time removed the oil from the barrel and cleaned it until it almost hurt the eyes to take a look down. The Colonel told me coldly that I was insolent. He awarded me eight days confined to barracks and added that I must pay for the cleaning of the rifle. I flushed red with anger at this injustice: I had a difficult job to restrain myself, bayonets or no bayonets, from slugging the Captain one on the jaw and kicking the table up over the Colonel.

I was now a defaulter or "on jankers" as the troops called it. Every time the bugle sounded the Defaulters' call, unless I was already on parade, I would have to answer my name at the

Guard-room and be marched off with the other defaulters to do some fatigue or other, in or about the barracks. Defaulters also had to parade every afternoon in full marching order under the Provost-Sergeant who for one whole hour marched them backwards and forwards over a space of ground no more than ten yards in length. The old "shoulder arms!" was not yet abolished, and carrying our rifles at the shoulder with bayonets fixed numbed the middle finger and made the right arm ache unbearably: the bayonet seemed to make the rifle twice as heavy. At the end of this hour, which was no joke, we had to show our kit, which we carried in our valises. If any man was short of any article he was tried for the crime on the following morning and generally given an additional three days confined to barracks. Between six o'clock in the evening and Last Post the bugle sounded the Defaulters' call every half an hour and at every call the Defaulters had to answer their names to the Sergeant of the Guard. They were not allowed in the Canteen at noon, though they could visit it between eight and nine at night, with an interruption at half past eight of going to the Guard-room to answer their names. The Defaulters concluded their day by falling in on staff parade outside the Guard-room at Last Post: this was when the Regimental Sergeant-Major took the reports of the Orderly Sergeants and the N.C.O.s who had been on duty in the barracks during the evening. During this eight days I was more fed up with the Army than I was during the whole of my soldiering. It was the injustice that made me feel so sick.

On the day I completed my sentence the Militia Captain left us and one of our own regimental officers, Lieutenant Maddocks, took charge of the Company for the time being. He proved himself as much of a gentleman as the Militiaman had been a pig. Before the Militiaman left he had ordered my rifle to be taken to the Armourer, and I was given another one in its place. Later I found out that my rifle had not been to the Armourer at all: it had been taken to the storeroom of the Company and put in oil with the other spare rifles – according to Musketry Regulations for the care of arms when not in use. We signed our accounts for the month an hour or two before we were paid out. As I was about to sign mine I noticed the sum due to me was a shilling less than what was due to the other men. The Colour-Sergeant reminded me of the shilling deducted for the cleaning of the rifle. I refused to sign. I said it was a damned robbery. But he threatened to make me a prisoner for using foul language to an N.C.O., and at the same time warned me that, if I did not sign, the Commanding Officer would have another bang at me. I was at last old soldier enough to realize that I was up against it. I signed. But I swore to myself that if ever I dropped across the Captain or Colour-Sergeant in civil life I would call them to budget over this affair.

I had been on the signalling course for six weeks when I was warned for a South African draft. After passing the medical officer I reported to the Signalling-Sergeant, who wanted to know where I had been. He was not aware until I told him that I was down for a draft and that I was proceeding to Blaina on a forty-eight hours' leave the following morning. He then began to swear and curse and said that I would see no Africa until I had completed my sixty days' course. I knew very well that once a man commenced a course he had to complete it, though the instructor could use his own judgment at any stage of the course and return any of the class to duty who either could not or would not learn. I badly wanted to go to Africa. I pleaded with the Sergeant to let me go. Our class had dwindled down to ten, the others having all proceeded to Africa. I pointed out to him that he had let them go without a murmur. "Yes, that's right," he replied, "and if you had been as dull as what they were I'd have let you go without a murmur too. But only you and Mills, of your regiment, are ever likely to make signallers, and in you two I shall have at least something to show that my work has not been in vain." He went to the Adjutant, who had my name struck off the draft. Another man took my place, and I cursed myself for having got on so well with the course. At the end the Sergeant recommended Mills and me for a further course of instruction at Aldershot; but we never went, because the details of the Royal Welch Fusiliers, who were now over six hundred strong, were put under orders to relieve the Devons at Jersey in the Channel Isles.

DRAFTED OVERSEAS

The old signaller, Mills, and I were sent over to Jersey six weeks in advance, under the charge of a lance corporal, to get the hang of the system of telegraphy in use there. For these six weeks we were attached to the Devons, who were stationed in the Fort at St. Helier's with a detachment of about one hundred men at St. Peter's, about six miles distant. I was sent to St. Peter's, where, instead of the proper telegraph instrument I had expected, I found a peculiar apparatus which worked according to what was called the Whitehead system. It consisted of two clocks, one a sending and the other a receiving clock. The sender had little buttons all around the outside of it and on the dial was the alphabet from A to Z and the numbers from one to zero; when a button was pressed a needle swung to the letter or number marked. The receiver had no buttons, but the needle swung around similarly to the letters or figures of the message; after the completion of a word it came to rest at a mark between zero and letter A. Each station was linked up with telegraph wires, but these were smaller than the usual ones and the clocks had to be charged electrically now and again. Messages could be sent and received extraordinarily fast; the drawback was that a message could not be written down while it was being received, so that every word had to be memorized. After I had been at the job three months I could receive a message of forty or fifty words and write it down practically word for word, figures and all. I worked from nine in the morning to noon and then from two to four. At St. Peter's I was nearly always finished by noon, as it was very rarely that an officer was to be found in barracks during the afternoon. I had a bunk of my own near the Orderly Room and I could stay out all night if I wanted to, so long as I was back in Orderly Room the next morning by nine.

I had the good fortune to pick up with a girl in a village not far from St. Helier's and we used to meet frequently all the months that I was in the island. There was no talk between us of marriage, as there was in the case of a number of our chaps who walked out with young ladies, preening themselves like peacocks beforehand. This was called square-pushing, a term which had its origin in the care that men took to get their knapsack to look properly square before parading in full order; but only one marriage resulted from all this effort, and then the man had to serve very nearly seven years in the East before he saw Jersey and his wife again. The people of St. Helier's were extremely respectable and prosperous owing to the trade in new potatoes and early tomatoes (with cheap manure from the sea-weed on the coast) and the summer tourist trade; and there were remarkably few prostitutes in the town, not more than half a dozen at the most. One of them was a very handsome girl with plenty of money which she made out of the summer visitors. Her name was French Annie. She would often treat a soldier who was hard up to a drink, a thing that I have never seen happen elsewhere. Not a single man was admitted to hospital with venereal disease during all the time we were stationed on the island, and that also was a remarkable thing.

I was doing only one hour's work a day; the rest I spent chiefly in reading. Although I detested school I had always been passionately fond of reading, and a Jersey Militia Officer who was attached to the Devons kindly lent me a number of novels, mostly historical ones, which was the sort I preferred. One thing that I had learned during my twelve months of soldiering was that the less work a man did in the Army, the more money he received – that is, if he was on a staff job like the one I was on now. Because I was doing next to nothing I was paid sixpence a day extra. At the end of the month an Engineer sergeant would come up to St. Peter's especially on purpose to reward me for my arduous work, and this fifteen shillings or fifteen-and-six always meant a good celebration for me in St. Helier's. At St. Peter's there was only a pub and a farm. Lily Langtry, the famous actress and friend to King Edward, lived close to Beaumont Station, which was the one I usually took to get into St. Helier's by the small railway which ran along the sea-front of the Bay of St. Aubyns. Her house was not a large one but very pretty and she was still a handsome and stately lady who had clearly been a dream of beauty when she was in the twenties. I got on excellently with the Devons and thought that Jersey was a grand place. It was almost a foreign country. A great many of the inhabitants were French-speaking and the island only recognized His Majesty as Duke of Normandy, not as King of England, which they regarded as a later and inferior title. There was no duty on spirits or tobacco; the dearest thing was beer, which was threepence a pint. I spent the happiest six months of my life in Jersey; it was just the same for me when the Royal Welch relieved the Devons, except that my last two months were spent in the Fort at St. Helier's.

Perched very high on the Fort was a Government visual-signalling station: there were men posted there with powerful telescopes, always on the look-out for any foreign warships that came within range. Whenever they spotted a warship of any description (and they were always French that they spotted) they would write a message giving the ship's name. The message was brought down to me to transmit to the Intelligence Department. Sometimes the messages were in cipher and they must have been important ones, because the cipher had to be repeated back to me to make sure that there was no mistake.

There was a kind of compulsory military service in the Channel Islands: every able-bodied male who had lived in Jersey a certain time had to join the Jersey Militia, which did about a month's training every year. Our chaps and the militia took part that summer in a sham battle on a piece of ground called Gory Common. (Jersey is so well-cultivated that there is not much space there for military manoeuvres.) I believe this battle had been arranged for the benefit of the aristocracy and well-to-do people of the island who were present in large numbers to witness it. It was a pukka comic-opera battle: thousands of blank cartridges were fired and the civilians cheered themselves hoarse when the charge was made. This was just about the time that the South African War came to an end. Peace was proclaimed on the last day of May 1902, and soon I had a letter from my Aunt to say that David had come safely through and was in the pink. Twelve months later, however, when I was abroad, news came that he had been invalided out of the Army with some complaint of the heart which had been brought on by the hardships of Mounted Infantry service on the veldt. He was granted the noble pension of tenpence a day for life. Later he married and removed from Blaina to Bedwas where he had a good job; he interested himself in local government work there and was soon elected a councillor, but the South African War had left its mark on him and he died before he was fifty, honoured and respected by all who knew him.

Several men had been trained as telegraph operators to take our places in case anything happened to us. I was very glad that this had been done when the word went round that a draft of one hundred and ninety men would be sent to India, as soon as the trooping season commenced, to reinforce the Second Battalion who were under orders to proceed there from Hong-Kong. My name was down for this draft and I went home on leave to Blaina. When I arrived there I found the people not nearly so interested in soldiers as what they had been while the war was in progress. My Aunt and Uncle made me very welcome and if I had not changed so much during my short service it would have been just like old times; though I missed my old buttie, who had moved down to the Rhondda. My cousin Evan was still working in the same pit as before. (He was still there when I returned after completing my service in India, and even when I returned after my four and a half years' service in the Great War. In fact,

43

he worked there for twenty-five years, until 1921, when the colliery failed to reopen after the Strike. Then he moved to Birmingham. Evan has had eight children and has been a grandfather for a few years: it is remarkable how different his life has turned out from mine, considering how close together we were as boys. Evan is also a staunch Labour man.) My Aunt was always a great believer in the grub stakes, and on the day I returned to duty she packed up enough of sandwiches for my train journey to have appeased the appetites of a dozen men. Her final advice to me was to have faith in God; my Uncle's was to have faith in myself and keep away from the damned women, who were the ruin of many a young man who went out to India.

During my leave the Royal Welch details left Jersey for Lichfield in Staffordshire, where I rejoined them. I well remember the night before the draft left for Southampton: it was a wild one. Most of the draft started mafficking, as we called it then, or making whoopee as the same business is called now. All the furniture in the Canteen suffered severely. There was fighting through and through and armed pickets were called out to restore order. We had a fine display of black eyes and cocked lips to show the following morning, but everybody seemed in good spirits when we marched to the railway station and one man who had the blackest eye of any, said that he had spent the most enjoyable evening and night of his life. Later, we had to pay for the damage that had been done in barracks: each man of the draft was stopped twelve shillings out of his accounts. There were many groans and curses over this, from guilty and innocent alike: there was not £100 worth of furniture to smash in the whole Canteen, and that is counting in the old piano which was so old and out of tune that it was next to impossible even to knock "Pop Goes the Weasel" out of it. No doubt the Canteen was provided with a new piano out of the damages that were paid by us in distant India, and no doubt there were some who had a good drink with the surplus money after the furniture had been repaired.

We embarked at Southampton on a troop-ship, the name of

which I have forgotten. There were close on two thousand troops on board and we reached Bombay towards the middle of November after a voyage of twenty-one days. I had never been on a long sea voyage before and I enjoyed every day of this one. I did not suffer with sea-sickness as many did: in fact I have never been sea-sick in the whole of my life. The food was good and the ship's bakers, who provided us with excellent bread, made money too by selling us penny buns. There were also sailors who came around during the day with large tins of sherbert and pails of fresh water; they charged a penny for a small tumbler of water with a teaspoonful of sherbert in it. This drink was called a Bombay fizzer. During the last ten days of the voyage they did an excellent trade with their fizzers and if they did not make a daily profit of thirty shillings to two pounds apiece they considered that trade was pretty bad.

Each man drew a hammock every evening and returned it the following morning. The hammocks were slung on hooks below deck, but after we passed Gibraltar we were allowed to sleep on the hurricane-deck and the forecastle. Many of us took advantage of this permission and unless a storm was raging I always chose the upper deck, in preference to trying to sleep in the heat below deck. But I had to move very quickly in the early morning if I did not want the swabbers who were cleaning the decks to turn the hose on me and wash me into the scuppers. Each unit in turn found a guard, with many sentries posted both below and above decks. The only parade we did during the day was about an hour's Swedish drill in the morning, but we occasionally paraded with lifebelts on for boat-drill. Most of the time was passed away in gambling. There were card-parties of Kitty-nap and Brag dotted here and there on the hurricane-deck and forecastle, but Under and Over, Crown and Anchor, and House were the most popular games. It was generally old sailors or ship's stewards when they were off duty who worked the Crown and Anchor boards. Two of these stewards, father and son, used to relieve each other on the job.

In case anyone is interested in a description of Crown and Anchor I shall give one here. The requisites are a board or sheet, three dice and a dice-cup. The sheet can be easily made and anything will do for a cup, but a good set of dice costs about a guinea. The whole outfit can be stowed away in the corner of a man's haversack. On each of the dice are six figures with corresponding highly coloured figures in six squares on the sheet. These are the figures with their nicknames:

Heart Transfixed	Puff and Dart
Diamond	Kimberley
Club	Shamrock
Spade	Grave-digger
Anchor	Mud-hook
Crown	Sergeant-major

The banker sits on the ground with legs wide apart and the sheet between them. The whole of his money is placed in two caps, one holding bank-notes, the other silver. He puts them quite close to him between his legs.

To attract players the banker shouts as follows: "Come on, come on, my lucky lads! Gather around the old man and see the game of Diddlum Buk. Here you are, my old sports, chuck it down thick and heavy. The more you put down, the more you pick up. You come here in rags and you drive away in a carriage and pair." After everyone around the sheet has backed his fancy, the banker shakes the three dice in the cup, which he turns upside down on the sheet, the dice remaining covered. If, say, the diamond and the heart transfixed have been poorly backed, the banker, before lifting the cup to display the dice, will encourage the backers to lay a little more on them: this is to square up his book. He will say: "What about the old Kimberley and Puff and Dart, my lads? They're the very two that should be backed. What wins the ladies' hearts? Diamonds, my lads, diamonds. Now then, is there any more for any more? Have you all done? Well, up she comes and-

46

what did I tell you? – one lucky Kimberley, one Puff and Dart and the grand old Sergeant-major! I went down on my knees begging you to back two of the three but you wouldn't listen to my advice … All paid, all weighed, and off we go again."

Even money is paid for figures that come up singly; two to one when they come up on two dice, and three to one when they come up on all three. A backer can put his stake on as many figures as he likes. If an equal stake is laid on each figure and three different figures turn up, the banker has nothing to payout. Some bankers put ten shillings or more on the centre of the sheet, to be won by the man who has the largest stake of the round on a figure that turns up on all three dice at the same time. This is done to incite men to lay a heavier stake on a figure than what they otherwise would. If this prize is won a similar amount is immediately put on the sheet again, the banker pretending that he is delighted that his customers are in luck. If more triplets turn up he stops giving money for nothing, but the odds are over two hundred to one against a triplet. A favourite system with backers who have plenty of money and are out to beat the banker is to back a certain figure and stick to it. They start with a small stake and double it every time they lose. If the original stake has been sixpence and the figure has not turned up for eleven rounds, over £50 will have to be staked on the twelfth bet; if it then turns up singly all that has been won is the original stake. To win any money with this system, doubles or triplets of the figure have to turn up at some time or other, and the odds against a double is eighteen to one each time. I have always considered this system a rotten one. The only time I ever won real money at this game I backed a different figure every time I made a bet; but as a system this is rotten too. I nearly always used it during my Army service, but with that single exception always rose up stony broke at the finish. All the men I knew who worked a Crown and Anchor sheet were rolling in wealth; the odds were considerably in their favour and they had to be very unlucky to go broke.

47

The trooping season lasted six months and the old steward told a few of us one day that for years he and his son had been winning five hundred pounds or more in the course of each season. He also revealed that more money was won from the time-expired men coming home than from the recruits going out.

While I am on the subject of games of chance, I may as well explain Under and Over, and House. The requisites for Under and Over are two dice, a sheet and a cup. The dice are ordinary ones numbered from one to six. The sheet is divided into three squares; one of the outside ones is marked "Under" and the other "Over"; the centre square has the number seven marked on it. The banker's remarks are similar to those of the Crown and Anchor banker's, and the dice are shaken and covered in the same manner. If the dice when uncovered show that the number of pips, added together, is under seven, "Under" Wins; if over, "Over" wins. If they add up to seven, the Lucky Old Seven, as it is called, wins. The banker pays even money on Under and on Over, but three to one on the Seven. This game was not so popular as Crown and Anchor, and before I left the service it was very rarely that I saw it played.

The requisites for a game of House are twenty-four cards, ninety small round wooden pips numbered from one to ninety and a small bag to shake them in; total cost about half-a-crown. The whole of this outfit also could be stowed away in a man's haversack. There are three lines on each card, the majority having five numbers on each line; although similar numbers appear on some of the cards they are so arranged that not one single line is the same as another in the whole pack of cards. It takes two men to work a game of House: one calls out the numbers, the other collects the money and issues the cards. Before the first game begins it is generally decided between workers and players whether a penny or twopence shall be paid for each card on the games played. The first man to cover a single line on his card and call out "House" is the winner. Out of every sixpence taken the workers deduct a penny for themselves. After the first game-which takes longer because of the players getting down and disputing together – the whole process of collecting the money, calling the numbers and covering a single line only takes four or five minutes. Pieces of bread, match-sticks or small stones are used for covering the numbers as they are called out.

Some workers I knew would not charge for the cards for the first game and gave eighteen pence out of their pockets to the winner – this was called a free house. Some would attract players by the power of their lungs and the quick way they called the numbers out. To draw a crowd they would shout: "Housee, housee, housee! House about and what about it? Roll up, my lucky lads and patronize the old firm! If you are lucky you may win enough for a good night's booze, and enough, besides, to spend the livelong night with the most charming White Pros. in Bengal (or wherever it might be) to roll up and take a card!" The caller sits on his backside, legs straight out in front of him. He shakes the pips in the bag, draws out a handful and shouts: "Look on! Eyes on!" As he calls out the numbers he lays each pip on the ground in front of him, making nine rows, so that it will be easy to check the card that "House" has been called upon. If House is not called on the first handful he remarks: "Another dip in the old bag and nobody sweating." "Been sweating a blasted week for one bloody number," immediately reply several voices in chorus. When House is called the card is handed to the caller, who checks it with the pips on the ground, which he calls out in a loud voice. Finding it correct he hands the card back saying: "House correct, pay the man his money." It was very rarely that a house was found not to be correct: it always occurred through the man innocently covering a number on his card which had not been called out. The number in dispute might be on the card of some other man who had been waiting for it to complete his line. Often before the caller had time to shout "House not correct" the majority of the men, thinking the game was over, had knocked all the pellets off their cards. The caller would then have to call every number that he had already called, over again; before doing so he would curse a little and invite his assistant to examine the man's ears. The players would also curse and tell the man to report sick and have his bloody ear-holes syringed out.

Good callers always called the nicknames for the following numbers:

No. 1 – Little Jimmy, or Kelly's Eye.

No. 11 – Legs Eleven. (The number resembles a pair of legs, and was given this extra syllable to distinguish it from Seven and avoid mistakes.)

No. 28 – The Old Brags. (The nickname of the 1st Battalion of the Gloucester Regiment, the 28th Foot.)

No. 44 – Open the Door.

No. 66 – Clickety-click.

No. 90 – Top of the House, or, Top of the Bleeding Bungalow.

One caller was so well versed in military history that as he called each number he would give the name or nickname of the regiment of the line that corresponded with it, from Number 1, the Royal Scots, or "Pontius Pilate's Bodyguard," onwards. For example, Number 9 was the "Holy Boys," the Norfolk Regiment, who once sold the Bibles given them by a pious old lady, before going overseas, to buy beer. Number 50 was the "Blind Half-Hundred" or "Dirty Half-Hundred," the Royal West Kents. Number 57 was the "Diehards," the Middlesex Regiment. Number 69, as I have already explained, was the "Ups and Downs," the 2nd Welsh Regiment, who started as a mixed battalion of old crocks and young recruits, then fought for some time as marines, and at the finish, after nearly two hundred years of service, were officially converted into Welshmen. Number 23 was our own regiment, the "Old Flash and a Dash."

During rests on the line of march I have often played this game for hours without getting up for anything. The game would sometimes continue all through the evening up to an hour before First Post, by candlelight. The last half dozen games were always Full House games, played at fourpence a card: the players had to cover all three lines on their cards before they could call House.

The workers considered that they had had a bad day if they did not make between thirty shillings and two pounds between them; but this would mean six or seven hours' continuous play. There were always men waiting ready to take the place of players who had to drop out of the game. Although all gambling was strictly prohibited, even the most regimental of the N.C.O.s in the Second Battalion always winked an eye at it. Most of them were fond of a gamble themselves and on the line of march everyone of them had a flutter now and then – with the exception of the Regimental Sergeant-Major and the Colour-Sergeants, who had their dignity to keep up.

THE LAND OF MILK AND HONEY

At Bombay we were taken ashore by tenders and stayed in the troop-sheds until the evening, when we left by train for Deolalie, which at this time was a depot for all troops arriving in the country and also for all time-expired men who were leaving it. The trooping season began in October and finished in March, so that time-expired men sent to Deolalie from their different units might have to wait for months before a troop-ship fetched them home. Moreover, if a man completed his seven years with the Colours on the 30th September he caught the last boat of that trooping season; but a man who completed his seven years on the 1st October would have to serve another year with his battalion and catch the first boat home the following trooping season. The time-expired men at Deolalie had no arms or equipment; they showed kit now and again and occasionally went on a route-march, but time hung heavily on their hands and in some cases men who had been exemplary soldiers got into serious trouble and were awarded terms of imprisonment before they were sent home. Others contracted venereal and had to go to hospital. The well-known saying among soldiers when speaking of a man who does queer things, "Oh, he's got the Doo-lally tap," originated, I think, in the peculiar way men behaved owing to the boredom of that camp. Before I was time-expired myself the custom of sending time-expired men to Deolalie was abolished: they were sent direct to the ports of embarkation, which in some cases meant weeks of travelling, but they got on the troop-ship the day they arrived at the port.

We reached Deolalie at daybreak the following morning and

were issued with ground-sheets, blue rugs and Indian kitbags. On the second morning a few of us strolled around the tents of the time-expired men. The majority of these had gone on a route-march but there were a few of them outside their tents and we got into conversation with one of them who had completed over twelve years' service, eleven of these in India. There was a native sweeping around the tent where we were conversing, and the old soldier ordered him on another job. The native replied in broken English that he would do it after he had finished his sweeping. The old soldier drove his fist into the native's stomach, shouting at the same time: "You black soor, when I order you to do a thing I expect it to be done at once." The native dropped to the ground, groaning, and the old soldier now launched out with his tongue in Hindoostani and although I did not understand the language I knew he was cursing the native to some order. The native stopped groaning and rose to his feet, shivering with fright: the tongue of the old soldier was evidently worse than his fist. He made several salaams in front of the old soldier and got on with the job he had been ordered to do.

The old soldier then said: "My God, it's scandalous the way things are going on in this country. The blasted natives are getting cheekier every day. Not so many years ago I would have half-killed that native, and if he had made a complaint afterwards and had marks to show, any decent Commanding Officer would have laughed at him and told him to clear off. Since old Curzon has been Viceroy things are different, you see. An order has been issued, which every soldier in India believes came from him, that Commanding Officers must severely punish men who are brought in front of them for ill-treating natives. We have to be very careful these days. If we punch them in the face they have marks to show, so we have to punch them in the body. Most of the natives on the Plains have enlarged spleens, and a good punch in the body hurts them more than what it would us. I expect you lads have got six or seven years to do in the country, and if you live to become time-expired you will have the same feeling towards natives as what I have. You will soon find out that the more you are down on them the better they will respect you. Treat them kindly and they will show you no respect at all. What is won by the sword must be kept by the sword, and it's the only law that will ever apply to this country. Old Curzon is no damned good, this country wants a Viceroy who will keep the bleeding natives down. If I had my way I'd give him the sack and recommend him for a job as a Sunday School teacher among the Eskimos around the North Pole."

It was drawing on towards Canteen-time and he enquired if we were fond of a drop of "neck-oil," which like "purge" was a nickname of beer. When we replied that we were, he exclaimed: "That's good! So long as you are in this country always have your drop of neck-oil and you'll live all the longer for it." While we were in the Canteen natives came around selling monkey-nuts. They were shouting: "Monkey-nut wallah, plenty good with the beer." Monkey-nuts were very cheap and most of the time-expired men were buying them and dipping them in salt to eat with their beer. The old soldier recommended them, saying that they made the neck-oil go down better. Before stop-tap we agreed with him and the ground around us was thick with monkey-nut shells. During our forty-five minutes in the Canteen he gave us some good advice, especially regarding the native prostitutes, and told us to mind that whatever we did we should never go with one of the numerous prostitutes that were always soliciting on the outskirts of camps or in the neighbourhood of barracks.

We left Deolalie on the evening of the following day and the old soldier wished us the best of luck. As the years rolled on I came to agree with him as to how the natives should be treated, and I still agree with him that what is won by the sword must be kept by the sword. During my first two years in the country I found that Lord Curzon was very much disliked by the rank and file of the Army, who all agreed that he was giving the natives too much rope. Another thing that added to his unpopularity was that his wife, Lady Curzon, was supposed to have said that the two ugliest things in India were the water-buffalo and the British private soldier. Every soldier in India at the time believed this story and very much resented it. A water-buffalo is larger than an English cow and a very ugly beast indeed. One of our chaps said that he would like to see the whole of the Battalion parade naked in front of Lady Curzon for inspection, with Lord Curzon also naked in the midst of them: for comparison, like a tadpole among gods. Our destination was Meerut, the city where the Indian Mutiny broke out. The Second Battalion, as we found out later, had left Hong-Kong and arrived at Calcutta the day before we reached Bombay, but the advance party that pitched the camp at Meerut for the Battalion was a draft that had arrived from the First Battalion in South Africa about ten days previously.

It took us twelve days to reach Meerut. We travelled by train at night, at a very slow pace, and stayed in rest-camps during the day. The rest-camp at Jhansi was about a mile and a half from the city, where plague was raging. The city and the bazaars were out of bounds for the troops. We stayed at this rest-camp two days. Early in the afternoon of the first day a party of us went down to a small stream to wash our feet. This stream ran through a valley about half a mile from the camp. Just after we arrived at the stream a native approached us with half a dozen girls walking in file behind him: they seemed to be all between fourteen and twenty years of age. He said that the girls were plenty clean and were from the brothel in the Suddar Bazaar where only the white sahibs visited. If we wanted to go with one of them we could have our pick, and it would only cost us six annas. I would not risk it myself, but a few of the party did. The word reached the camp of what was in progress and in less than no time a large number of men appeared on the scene. The native took the money while the girls did the work. The stream was very handy; it enabled the girls to wash themselves and they did not mind in the least who was looking at them while they were doing this. By sundown the native had a decent bit of money tucked away in his loin-cloth.

An old soldier, who belonged to the draft of another regiment and who had served in India before, was determined to rob him of this hoard. He was called "The Soaker" and he had a reputation as a beer-shifter; he also had a reputation for borrowing money which he was never known to pay back. During the day he had lost the few rupees in his possession at the Crown and Anchor board, and was determined to get money from somewhere in order to enjoy his nightly issue of purge at the Canteen. Just after sundown the native was preparing to leave the valley with the girls when the Soaker said to a few men around him: "By God, boys, it makes my heart bleed to see that black bastard with the money which those girls have earned. He'll only give them just enough out of it to buy themselves food over tomorrow, and the rest of it he'll keep for himself. I've a good mind to relieve him of his ill-gotten gains. I'll give the girls eight annas apiece out of it, and that will be a hell of a lot more than what he would ever think of giving them." He approached the unsuspecting native, who soon lay groaning on the ground from a hard blow to the stomach. The Soaker had little difficulty in taking the money off him, but when he had done this the girls started wailing and weeping and shouting, "No pice, Sahib, no connor, Sahib." hey meant that if the native had no money left him, they would have to go without food. The native suddenly recovered and rising to his feet shouted at the top of his voice, "Loose-wallah," which was Hindoostani for "thief." The Soaker spun him around and booted him along, at the same time cursing him in Hindoostani. The native broke into a run and soon disappeared. The Soaker gave the girls eight annas each, telling them that if they did not quickly clear off he would take it back off them. In a short space of time they also had disappeared from view. The valley was out of sight of the camp, and the Soaker said in a righteous manner that he had done the girls a good turn. If a couple of the military police had appeared on the scene at some time during the day they would have said nothing until the native was leaving with the girls; then they would have pounced on him, robbed him of his money, given him a damned good hiding and driven the girls away without a pice-piece.

We had changed our English money into Indian currency with the Parsee money-changers at Bombay. The Indian currency was easy to get onto: a rupee was worth one shilling and fourpence, an anna one penny, and a pice one farthing. We arrived at Meerut. The Battalion had arrived the day before us and was quartered in large marquee-tents which had been pitched not far from the cemetery. They struck me as a very fine, tough lot of men. Some seven hundred men out of the thousand-odd were Cockneys and Midlanders in equal number, not three hundred being proper Welshmen. About this time the Cockneys and Welsh grew fewer, and the Midlanders more numerous, until in 1914 the Battalion was sometimes jokingly known as the Birmingham Fusiliers. Birmingham men, however, made the best soldiers in the Army, and on Taffy Day, as we called St. David's Day, they proved themselves as good Welshmen as could be desired. (We were then issued out with large leeks which we wore in our caps on parade, and also when walking out; if we happened to be in a station with other troops and any sneering remark was made about the leek it was immediately paid with a hard blow for the honour of Wales. And they never made a fuss when called on to obey the old Taffy Day custom of standing on the Barrack-room table and eating a leek; for which their only compensation was a pint of beer to wash it down.)

One of the largest rifle-ranges in India was on the side of the main road. About five hundred yards away on the other side of the road was a smaller rifle-range which had a couple of straw ricks behind it – both ranges were close to the camp. I was posted to C Company; I met quite a number of men who had enlisted about the same time as myself and had been sent to Africa after I had left Plymouth. Two drafts had been sent; the first arrived at Cape Town about two months before peace was proclaimed. As they had travelled through three different colonies to reach the Battalion, which was now in block-houses – the war had only been won by the slow method of pushing the Boers back with a long line of block-house and wire – they were awarded a medal and three colony-bars. The second and last draft to Africa were always known as the cease-fire draft; they were still a week's sail from Capetown when peace was proclaimed. If they had arrived at Capetown only twenty-four hours before peace was proclaimed, they would have been entitled to a medal and perhaps one colony bar. Some men of this draft used to complain that they had been cheated out of their medals: they seemed to think that they had been largely responsible for bringing the war to a close. No doubt the Boer leaders had the wind up as soon as they knew this draft was on its way and quickly made up their minds to sue for peace.

Although it was winter the days were nice and warm, but the nights were bitterly cold. It was an intense dry cold which seemed to penetrate right through a man's body. We laid our waterproof sheets on the ground with blue rugs on the top of them and got in between the blankets. The cold caused quite a number of men to wet their beds regularly: this was called "having a benny." And when we woke up in the morning the ground was heaving with white ants which ate holes in groundsheets, blue rugs and everything else: they are supposed to be the most intelligent and also the most destructive members of the ant family. Many of us bought native beds called charpoys: they were made with four wooden posts and a bamboo cross-piece at each end. Cords were stretched between the cross-pieces to act as a spring. These charpoys only cost eight annas each, but after they had been in use for a month or two the cords would sag in the middle, so that when a man got into bed his backside would be about an inch from the ground. Also, the white ants would play havoc with the posts, which were soon full of holes where they had been at work: three months was a decent life for a charpoy. The Indian kit-bags were ant-proof, if one laced them up tightly, and it was rarely that our clothes got damaged. But if we had to sleep on the ground for more than a night or two, our thick blue rugs were soon ridded with holes.

The Battalion had its own dairy, bakery and Regimental Bazaar. This Bazaar contained the shops and dwellings of the natives who had attached themselves to the Battalion. The dairy provided the butter and milk that was used by the Battalion; the bakery, bread; and in the Regimental Bazaar everything was sold from contraceptives to bicycles. Natives came around the camp before breakfast and before tea, carrying trays with small pats of butter on them and also cans of milk. They shouted, "Mucking Wallah, Dood Wallah," which were the Hindoostani names for butter and milk. A pat of butter which weighed about an ounce cost one anna. A pice worth of milk was sufficient for the tea. The butter was excellent, but it was very dear considering how cheap other articles of food were. Under a large tree which was called "the Ration Stand" natives arrived every morning selling fresh meat, bacon, eggs, pork-sausages and vegetables. Everything was dirt cheap: best beef-steak one anna a pound, mutton-chops six pice a pound, best country bacon three annas and two pice a pound, pork-sausages three annas a pound, eggs were three or four for an anna. I now understood what the old soldier in the pit at Blaina had meant when he told me that India was a land of milk and honey. If a man felt like having a chicken for his dinner he never paid more than four annas for it. But there were times when a man owing to bad luck at Crown and Anchor, or a stoppage of pay, did not have the money to buy even a pice-worth of milk and had to subsist on his rations. The only difference between rations in India and rations at home was that here a soldier was allowed one pound of meat a day instead of three-quarters of a pound. We also did not pay threepence a day messing, so there was no mess-book for a corporal to shark and our pay was one rupee a day. From our meat rations a small steak was cut for each man's breakfast. These steaks were called khaki patches, and a man's jaws would ache for hours after he had masticated one of them. There was not much fat on any of the cattle that were killed on the Plains, but the cattle issued to the troops did not have enough of fat on their kidneys, it was commonly said, to fry the liver of a mosquito.

It was only during the winter that bacon was sold on the Plains, and one old man, who was called the Bacon-wallah, was always an early arrival under the large tree. He had three natives with him who carried his stuff and worked under his supervision; they seemed to be in mortal dread of him, as were all the other natives who stood at the Ration Stand. He was a shrivelled-up old chap about five feet six in height and when I first met him I could not tell whether he was a white man, a half-caste or a native. But it turned out he was white. He smoked a native pipe called a hookah or hubble-bubble: it held about an ounce of tobacco and he would sit on his haunches like a native while he was smoking it. It was common to see half a dozen natives in a circle, smoking and gossiping; they sat on their haunches with one hubble-bubble between them, from which each man took a few whiffs before passing it on to the next man. They smoked all kinds of stuff, including charcoal and live coke, but the old Bacon-wallah smoked our tobacco, which was very cheap. At this time there were no duties on tobacco and cigarettes, and best plug-tobacco only cost one rupee a pound.

I became very friendly with the old chap, who was an old British soldier who had served under the East India Company, or John Company as he called it. He was not sure of his correct age but thought he was knocking a hole into ninety. He once asked me when I had joined the Army. I replied, that it was the year Queen Victoria died. He smiled and said that he had enlisted in 1837, the year Queen Victoria was crowned. After twelve months' service at home he had been sent to India and had been nineteen years in the country when the Mutiny broke out. He had taken an active part in the fighting around Meerut and I was always interested in his yarns of the Mutiny.

One morning he looked up at the tree under which we were conversing and said: "Well, youngster, it's nice and peaceful around this tree, but in the early days of the Mutiny it was not so nice and peaceful around it. Many a morning before breakfast I was present here and assisted to hang a dozen or more of the rebel Pandies on the branches. I got so used to it that I found on the mornings that I was not lending a hand to hang a few of them that I could not relish my breakfast. You can take it from me, youngster, that it gives a man a wonderful appetite for his breakfast to assist at turning-off a dozen or more rebels. Ay, the more Pandies that were hanged, the larger the breakfast I ate. After a time the execution of Pandies by hanging was abolished: we used to tie them to the muzzle of a cannon and blow them to hell and back. Most of them held the religious belief that if they were blown to pieces their souls would be blown to pieces as well, which made it impossible for them to have another existence. This method of executing them did more to quell the Mutiny than anything else. They laughed at hanging, the bolder ones of them, because they believed they would have another existence; but they squealed like rats when they were tied to the cannon's mouth.

"After the Mutiny was over I left the service with a pension of a shilling a day for life. I had been so long in the country that I decided to stay in it, and quite a number of us old John Company soldiers who were entitled to a pension did the same. Quarters were found for us in the cantonments at Lucknow, and we managed all right on our bob a day, as victuals were twice as cheap in those days as what they are now. For six or seven months of the year I wandered about the country, visiting different military stations. I generally stayed a few days with each battalion and when I left a collection was made for me. After living like this for about twenty years I thought it was about time to settle down and take unto myself a wife. I must have been about sixty at the time. I married the daughter of a couple of half-castes. She was thirty years younger than me and her parents, who had a bit of money, gave her a dowry of a thousand rupees. With this money I started a piggery in a small way, which soon got much larger. I have now made enough of money to retire on, but somehow I can't, and I love to be under this old tree before breakfast in the morning."

This game old chap had taken part in many wars in India

before the Mutiny; and he must have been a sperky old sinner to have taken a wife, at sixty years of age, who had borne him four children, all now grown up. There were a few more of these old John Company soldiers left who still wandered round the country for six or seven months of the year in exactly the same way as this old chap had done. When we were stationed at Agra one of them visited the Battalion every winter. He was still a strong and powerful man for his age and always wore a breastful of medal-ribbons. He had fought against the Sikhs in 1845, was one of the relieving force at Lucknow in 1857, had taken part in the Chinese War of 1860 and also fought in Abyssinia. He became highly indignant when I asked him one day whether the ribbon which he said was the Abyssinian ribbon was for the campaign of 1868, under Lord Napier. He snorted and puffed out his cheeks and said: "1868, my lad! Whom do you think you're talking to? I fought in Abyssinia at a time when the backsides of the men who were destined to fight there in 1868 were no bigger than shirt-buttons." I think it was for 1847 that he said he won his medal, but I can find no record of any British campaign in Abyssinia at that date. Perhaps he was fighting against the Emperor Theodore (or Kassa as he was then called) for Ras Ali of Northern Abyssinia, who had British support: but Ras Ali was always defeated.

This Abyssinian warrior never stayed more than three days and a collection was always made for him before he departed. The amount of beer he could swallow was amazing and the dinner he could shift would have done credit to a recruit at the Depot. The last time he visited the Battalion he stayed in the Signallers' room. He told us that when he enlisted in the Army a soldier's pay was only a halfpenny a day. Although the pay was so small he considered that the soldiers under the old John Company before the Mutiny were far better off than what we were. They were generally engaged on active service, from which they derived much loot. After a little war was over they would have the times of their lives and generally by the time they were broke they would be sent to quell another rising in some place or other. I was having a drink with him one evening at the Canteen when he said: "Sonny, the soldiers of the old John Company drank rum and not shark's p—s. This rotten stuff will wash a man's kidneys away before it will make him drunk. In my old days it was a common sight by stop-tap to see practically every man in the Canteen as drunk as rolling f—ts: yet if they had not been put in clink meanwhile they would all wake up in the morning as happy as larks." Both he and Baconwallah, who also liked his drop of purge, said that the Army was by no manner of means what it used to be, and that the country was fast going to the dogs, by the way some of the natives were now strutting about. If they had had their way they would have hanged or bayonetted any native caught playing at being the equal of a white man. In their old days they used to half-murder any native who approached them without salaaming thrice to the ground.

At every meal-time there were thousands of kite-hawks and crows flying around the camp. The kite-hawks looked like small eagles. If a man who was up on a high building dropped a bit of meat, a hovering kite-hawk would drop like a stone and catch it in its claws before it reached the ground. Before we came to know these birds many a man lost his breakfast as he was carrying it from the camp fire to his tent. A kite-hawk swooping down would knock the plate flying, and down would go the khaki patch on the ground, where it would be left for the raider to eat, being covered with sand. The only way to keep these birds away when carrying a plate of food was to keep waving one hand to and fro over it until you landed safely in your tent. But I never in the whole of my life saw cheekier birds than the Indian crows. Not content with the tactics of the kite-hawks, they would come right into the tents, pinch anything in the way of grub-stakes that they could find and fly away with it. Each man's folded blankets with his kit-bag on top of the pile, were laid in line outside the tents each morning, and the tents swept out; if a man bought a pat of butter for his breakfast and laid it on his kit-bag he would lose it the very instant he turned his back. A crow would hop along and pinch it, and adding insult to injury would from a distance of a few yards caw with devilish glee at his victim – but it was more like a quack than a caw. The natives believed that when a British soldier died his soul entered the body of a newborn crow, which made the crows so cunning and daring.

The Battalion paraded for Church Parade every Sunday morning. Roman Catholics marched to their own place of worship, and if there was a place of worship for the Methodists they did the same; if there wasn't they danced for joy. Ninety-five per cent of the Battalion heartily detested Church Parade and would do anything in reason to get out of it. Over ninety per cent of the Battalion were nominally Church of England men and we marched to Church armed to the teeth. Ever since the Mutiny it had been the custom for troops to parade for Church with rifle, belt, pouch and side-arms and each man was issued with forty rounds of ball-ammunition. The idea was that if a mutiny broke out while the troops were at church they could move off at once to quell it. I always thought this was a good idea, but I don't know what the One Above thought of it: I expect He did many a grin when He looked down over Northern India on His armed worshippers. As we filed into the pews we fixed our rifles in the sockets of the hymn-book rack in front of us.

One Sunday the Chaplain was very long-winded over his sermon, and the troops got a bit restless. The man next to me whispered: "Well, Dick, I'm about fed up listening to that old bastard's bull; if he doesn't pack up shortly the Canteen will be closed by the time we arrive back in camp." I whispered back that he was no more fed up than what I was; and judging by the way the troops were coughing they were all of the same mind with us. The Chaplain continued his discourse and in a short space of time the coughing was general-high tenor, baritone and deep bass coughs coming from all parts of the church. One man in front of me dug up a series of whoops that sounded like a cracked foghorn. The Chaplain paused in his sermon for a few minutes until the coughing subsided; frowning and holding up his hand for silence. Then he said: "I know very well what's in your mind, men, but the more you cough the longer I shall preach." He kept his word and by the time we arrived back in camp the Canteen was closed and dinner was being served. If the prayers that were sent up had been answered that Chaplain would have been dancing a two-step in Hell before the next Church Parade arrived.

The following Sunday we were warned that any man who coughed in Church would be made a prisoner and put in the guard-room when we arrived back in camp. Two men of my company suffered with bad chests, which always troubled them during church service. On this occasion they did their best to refrain from coughing, and the result was that when they did let the cough go it came out louder and more prolonged than ever. The Chaplain ceased his mumblings and mutterings and eyed the men fiercely for a few minutes. Quite a number of other men now took advantage of the pause to clear their own throats, and the Canteen was again closed by the time we reached camp. The two men were accused of being the ringleaders of the disorder; when we got back to camp they were put in the Guard-room, where they would have to stay until the following morning, when they would be tried for their blasphemous crime. Both of them had fought in the early part of the South African campaign before being sent to China. No doubt their case was similar to that of my cousin David, and their chest trouble was due to the hardships of campaign on the veldt. They had not been in the Guard-room long before they decided to go sick. Early each morning the Medical Officer examined the men that had reported sick, so these two had to have a special sick-report made out for them. After the orderly corporal of their company had made one out they were marched to hospital with an escort of the guard. The medical officer who examined them there happened to be a better Christian than many a parson. After hearing their story he admitted them both to hospital with bronchitis and kept them for ten days before they were discharged. They could not now be tried for their crime, because their being admitted to hospital had given their coughs an official standing. This was a good turn done to the whole Battalion as well, for I can't remember another man ever being put in the Guard-room for coughing in Church.

The uniform we wore in India was somewhat different from the home-service issue. On the Plains in the winter we wore the thin, fine Indian khaki by day, and red in the evening: I think that a suit of red was supposed to last us two years. The Indian red jacket was lighter than the home-service one; and we were not issued with red tunics. During the heat of the summer on the Plains we always wore white on parades – the black flash was very conspicuous on it – but as the only parades then were Church Parade and funerals this was seldom. In the Hills during these months we wore the Indian khaki.

LOOSE-WALLAHS

The city of Meerut was about three miles from our camp but it was out of bounds for British troops. I never found out the reason for this, but it was generally believed that if a white soldier entered it he would never leave it alive. The Meerut garrison consisted of ourselves, a battalion of the Rifle Brigade, the 15th Hussars, some native infantry battalions and a native cavalry regiment. We became very friendly with the 15th Hussars and this friendship lasted a long time: years later when we were at stations not far apart friendly visits were exchanged between the two regiments. There was a fine polo-ground near by where I watched many a game; there are more thrills in watching polo than a football or cricket match ever provided. But good polo could only be played with good ponies and these could cost up to £100 or £200 each. As ponies were changed after each chukka, which only lasted a few minutes, a player had to have quite a string of them. Only wealthy men could afford to play. A good pony would turn like a cat on the ball and often seemed to know more about the game than what its owner did. Some of our officers were first-class players but the finest polo-team I have ever seen was the one that represented the 15th Hussars at this tune. Two of them, Bingham and Barrett (I forget their ranks), were with the English team that visited America just before the War and won most of its matches. The Hussars lost over sixty men from enteric fever this winter, and had to go under canvas for a time while their barracks were fumigated. But the barracks must have been very badly built as the following winter they lost a decent number of men from the same complaint.

Not far from our camp a number of cavalry Cadets were

encamped – the sons of native Princes and Rajahs. A sergeant major of the Hussars had a well-paid staff-job, putting them through their drill. Each Cadet had a splendid tent all to himself and a suite of servants to wait on him, and a stable of magnificent horses. They wore golden spurs and the handles of their swords were made of gold. They did a couple of hours' drill in the morning, and spent the rest of the day practicing polo or tent-pegging at which most of them were very clever. One morning a few of us were watching them going through their cavalry drill, which they concluded by charging an imaginary enemy. The question arose whether they would have been equally as brave if they had been charging a real one. One old soldier named Carr, who had served in India many years before with the First Battalion, said that an equal number of British Cavalry mounted on donkeys would have cut them to mincemeat if they had clashed in a real fight. He added that if they had to charge only half their number in infantry, nobody would be able to see their horses' tails for dust the moment after one volley had been fired at them. In conjunction with the remainder of the garrison we carried out some long and tiring field-days, which one day took a section of us through an evil-smelling village built on the banks of a stream. There we saw half a dozen pigs with their snouts buried in the bodies of three dead natives; by their grunts of satisfaction they were enjoying their meal. I had eaten some country bacon for my breakfast that morning, but I never ate any more bacon or any pork-sausages either so long as I was abroad.

The Battalion was short of signallers and did not have enough of men to take part in the annual signalling test. A class of thirty men was started. Although I had been through a course at home I was a long way removed from being an expert signaller so I began with this class in the same way as a new beginner. Only five of us became signallers out of this class, which included a man of my own company, called the Prayer-wallah, who was my best pal at this time. Indian regulations required six regimental and twelve supernumerary signallers of each battalion to take part in the annual test. An officer from the School of Signalling visited each battalion and the test was carried out under his supervision. He handed over the messages that were to be used in the test and took them away with him after the test was over. Messages were sent by heliograph, lamp, flag and semaphore; and points were allowed for accuracy and speed. After all battalions had completed their tests a list was made out showing the number of points each battalion had made, and what their position was on the list. The regimental signallers of the first three battalions on the list received six rupees a month extra pay, those of the next six battalions three rupees eight annas a month, and those below the first nine two rupees a month. The supernumeraries received nothing; they had to wait their turn to became regimental signallers which in some battalions meant waiting for years. There was keen competition among the battalions in this annual test; after the first, in which we did badly, we were generally in the first nine.

A year before I had completed my time a new test was introduced which allowed a certain number of points for a message that was read correctly in a given time; if the same message had been read correctly in half the time no extra points were allowed for it as they had been in the old tests. With this new test no extra pay was allowed for being in the first nine on the list: all regimental signallers alike were paid two rupees a month extra pay. This knocked all the competitive spirit out of the tests and greatly slowed down visual signalling. In the time of the old tests a man was not considered a good regimental signaller unless he could read and send accurately sixteen words a minute on heliograph and lamp. It took a signaler from twelve to eighteen months' continual practising, after having first passed an annual test, to work up to this rate. There was a great difference between individual men in their capacity to send on the helio, and in their capacity to read. Some men could read much faster than others, and yet if they tried to send at the same rate, it would be quite unreadable.

69

Every year we started a new class of signallers and always kept half a dozen on the job; this was done to replace men who died or who were going away time-expired. All the signalers kept together except on the line of march, when the supernumeraries were returned to their companies for the duration of the march. Only the six regimentals were not separated: they always marched in front of the Battalion.

There were eight companies in the Battalion and each company in its turn did a month's duty at Delhi Fort, which was about forty-five miles from Meerut. The river ran around the fort; inside were palaces and other beautiful buildings of the old Mogul Emperors. The fort was healthy enough for nine months of the year, but after the monsoon had broken it was a very unhealthy place for a time. Hardly a man who was stationed there during the months of August, September and October escaped an attack of malarial fever, which I heard a doctor say was of a more malignant type than the malaria contracted elsewhere. During these months the stench from the low-lying river, which was now swollen by the rains of the monsoon, was enough to knock a horse down, and the mosquitos were of an awful ferocity.

The Pathans were born loose-wallahs and although they belonged to the hill-tribes of the North-West Frontier were to be met with in every city, village or bazaar in Northern India. They were tall, strong, muscular men and made a queer contrast with the natives of the Plains who were extremely skinny looking objects, the majority having legs and arms like runaway matches. A Pathan's main ambition in life was to steal a service rifle and get it back safely across the frontier. The frontier tribes found it easy to procure small-arm ammunition, but up-to-date rifles were a problem. A service rifle at this time cost about three pounds; if a Pathan stole one and got it safely across the frontier he could easily sell it, if he chose, for twenty-five pounds. With this money he could build a native house, buy a plot of land and three or four wives, and be comfortably well off. Although three-fifths of the Battalion had served in China or Africa there were hardly twenty men, officers included, who had served in India before. Before I was transferred to the Signallers' tents I stayed in the same tent as Carr, who was daily warning us that one morning we would wake up and find our rifles missing. Before getting in between the blankets at night he pulled the bolt out of his rifle and put it in his kit-bag, which he used as a pillow. A rifle was no good without a bolt, but he took the additional precaution of always sleeping with his leg thrust through the sling of his rifle. He did his best to get the others in the tent to follow his example but they only laughed at him. They said that he was suffering with the Doo-lally tap. When I was persuaded to follow his example they said I had caught the Doo-lally unusually early in my Indian soldiering. The Colonel also was aware of the danger from rifle-thieves. He applied for locks and chains so that the rifles in each tent could be lashed to the tent-poles, but these were not forthcoming. I was told that he then wrote to Brigade Headquarters, informing them that unless they were supplied he would not hold his men responsible for the theft of their rifles.

The lines of the camp ran from Companies A to H. On one flank was the quarter-guard with two sentries patrolling the same beat; on the other flank a small guard of an N.C.O. and three men was found during the night. They mounted at six in the evening and were dismissed twelve hours later. This guard, which had a single sentry patrolling the beat, was not far from the fringe of the jungle, out of which jackals and hyenas came sneaking during the night. Most of the men much preferred the twenty-four hours on quarter-guard to the twelve hours on this other lonely one. To hear a pack of jackals howling in unison is enough to curdle the blood of any man who is not well used to the noise. The hyenas and jackals scrounged around the camp at night, picking up what garbage they could find. Many a night I have been awakened by a pack of jackals scuttling through the camp. Another animal whose cry at night we soon learned to recognize was the cheetah, a sort of small spotted leopard. The cheetah, which is very fleet of foot, would hang around the camp in the hope of getting a dog for his supper. Many of the men owned dogs and they would never bark once they had smelt a cheetah, but used to shiver with fright and curl up as close to their masters as they possibly could. If a gang of Pathans were after rifles they always smeared their bodies with cheetah-grease before starting out on their night raids. They moved swiftly and noiselessly and generally did their work about an hour before dawn. The cheetah-grease served two purposes: the dogs scenting it would think a cheetah was close at hand, and if a man woke up and made a grab at one of these half-naked rifle-thieves his fingers would slip on the grease and he would perhaps get knifed for his pains.

One afternoon, early in January 1903, half a dozen Pathans arrived in the camp with performing bears, goats and monkeys. They were well-trained animals and could do some surprising tricks, especially the goats. Somehow one does not think of goats as performing animals. They gave a show outside the lines of A Company and quite a number of us gathered around to witness it. Of course there was the usual collection after the performance but it only realized about two rupees and the Pathans had to be contented with that. Carr, who did not like the look of them, said that if they were not loose-wallahs themselves they were in some way connected with loose-wallahs and had only visited the camp to see what they could fish out. There were all sorts of natives who visited the camp – fakirs, jugglers, fortune-tellers, snake-charmers, and natives who called out "Nail-wallah" as they went up and down the lines of the tents. A nail-wallah only charged one anna for manicuring nails, cutting corns, removing in-growing toe-nails, removing tartar from the teeth and cleaning wax out of the ears. I always thought these wallahs really earned their money, but if one of them in removing a man's in-growing toe-nail happened to cut into the quick he was generally paid only with a kick and a curse. A fortnight before the Pathans visited the camp with their performing animals a new nail-wallah, also a Pathan, made his daily round of the camp during the afternoons. He proved an expert at his job and soon captured the trade of the other nail-wallahs.

Each Company found the men for guard in their turns. Quarter- and flank-guard mounted every morning at nine, but after being inspected the flank-guard on the fringe of the jungle was dismissed until the evening when they came on duty again. Men who had been warned for guard would put in many hours cleaning and polishing in the course of the afternoon on the day before they mounted, so that they would be spick and span in everything on the following morning. It became the custom of the men mounting guard on the following morning to hang their rifles in the loops of the tent; the others mostly put theirs in the flaps overhead. It may have been a coincidence, but in the light of what happened afterwards it looked mighty suspicious that, on the afternoon that the Pathans appeared with their performing animals, A Company was finding the guard for the following day.

Just as dawn was breaking the camp was in an uproar. The news soon flashed around that nine rifles had been stolen from one of the tents in A Company. Most of the men who were mounting guard that morning lived in this tent and one of them, waking up at the first signs of dawn, saw the form of a native disappear through the door of the tent. It was still dark inside the tent, and for a moment he thought that one of the native boys who were engaged by the men to clean their boots and straps had arrived a little earlier than usual, and had entered the tent to take some of the boots outside. He got up and looked outside and, not seeing a cleaning-boy about, struck a match. The light showed him that there was not a single rifle hanging in the loops of the tent. He aroused the others and hastily slipping on his trousers and boots rushed out of the tent shouting "loose-wallah!" at the top of his voice. He spotted about half a dozen natives running for all they were worth in the direction of the straw-ricks behind the small rifle-range. His shouting attracted the attention of the flank-guard sentry, who spotted the thieves when they were about four hundred yards off, not far from the ricks. He brought his rifle to the aim and opened rapid fire at them. In his excitement he forgot to raise his sights, but even if he had done so he would have been very lucky to have hit one of them in the dim light of dawn. They got behind the ricks and disappeared into the jungle.

A section of men dressed hastily and gave chase, but although they advanced a considerable distance in skirmishing order after they had passed the ricks they discovered no trace of the thieves. The civilian police soon arrived on the scene with a party of expert native trackers, but before the day was over these had to admit themselves beaten. If the ground had opened out and swallowed the loose-wallahs they could not have vanished more completely than what they had. A white police-superintendent said that the thieves were bound to be caught in a few days, because messages had been sent all over Northern India and the whole of the police department would be on the look-out for them. But the days became weeks and the rifles were still missing. The men who had lost them were not punished, however.

Seven months later at the mouth of the Khyber Pass a family of Pathans with a bullock-cart were being questioned by the police, before being allowed to enter it. Their cart, which was half full of stuff bought during their visit to the Plains, was searched in the customary way. Among the hoard was a closed coffin. One of the Pathans told the police that the coffin contained the corpse of his grandmother, and that he and his uncles and cousins who were accompanying him were carrying out her last dying wish: she had asked particularly that she should be buried where she was born. The lid of the coffin was unscrewed and sure enough the corpse of an old dame was there all right. Nothing was found on the men or in the cart that the police could have taken exception to, but the officer in charge was not satisfied with the search that his men had made. He decided to make one himself. First he ordered the Pathans to stand to one side and hold their arms above their heads, while his men kept them covered with their rifles. Then he bundled everything out of the cart, including the coffin. He found nothing, but it struck him that the old dame weighed a good deal more dead than what she must have weighed alive. He pulled her out and began tapping the bottom of the coffin. It sounded wrong. He now turned the coffin upside-down and knocked the bottom in; and there, in the false bottom, were the nine rifles that had been stolen from us at Meerut. They still had the regimental initials and their own numbers on the small brass plates on the stocks.

The loose-wallahs had travelled five hundred miles or more from Meerut to Peshawar and must have been remarkably clever to have evaded the police on the way. Once inside the Pass they would have been safe. It appeared that the corpse was related to one of the men, and no doubt he and the gang came to the conclusion that she had died at an opportune moment for their attempt to got into the Pass. After the rifles were stolen the new Nail-wallah was never again seen by the Battalion. During the fortnight he had been coming around the camp he had evidently used his eyes, ears and brains to some purpose. But if he was with the party at the Pass he would soon have the pleasure of cutting Allah's own toe-nails and cleaning out the wax from the Prophet's ears, for rifle loose-wallahs did not live long after they were caught. The Pathans with the performing animals were found in one of the bazaars, but although they were questioned and all their belongings searched the police were unable to prove that they were in any way connected with the theft of the rifles. We were stationed at Chakrata at the time and sent an escort to Peshawar to bring the nine rifles back.

In addition to locks and chains the Colonel had also applied for Snider rifles which were issued to each battalion in India for the use of sentries on guard at night. They were an out-of-date rifle, the first breech-loading guns issued to the British Army. They had been in use between 1866 and 1876. But while a sentry at night would be extremely lucky if he hit anything he fired at with the Lee-Metford rifle, which was the Service rifle at this time, with a Snider he would have an excellent chance of making a hit – buckshot cartridges were issued with them which, when fired, spread over the object aimed at. A week after the rifles were stolen the Sniders arrived, and were always kept in the Guard-room or Guard-tent for use by sentries between dusk and dawn: they had long thin bayonets fixed on them. The Regimental police now found two flying sentries who patrolled the camp at night, armed with Sniders. A lot of men were suffering from ground colic. Some of them were carried to hospital doubled up with gripings, others who were not too bad with it remained in their tents, except for frequent rapid journeys to and from the latrines. At night the poor devils went in terror of being spotted by a flying sentry and not answering his challenge quick enough: a group of buckshot in their backsides might further increase their troubles.

As the locks and chains had not arrived, orders were issued that each tent had to have a pit dug in the middle of it, large enough to hold the rifles of the occupants. After these pits were dug, tent-bags were laid at the bottom. The rifles were put in the pits every night before Last Post, with more tent-bags laid above them. Each man took his turn in sleeping on top of them during the night. Carr and I strongly objected to putting our rifles in the pit, but the N.C.O. in charge of the tent, who was not a bad chap, said that orders were orders, and that he had to see that every rifle in the tent was in the pit. Ours had to go in as well. The first man in the tent to sleep on the rifles was nervous at the best of times. Carr had made him more so by relating all the most bloodthirsty yarns about loose-wallahs he knew; saying he hoped that he wouldn't awake in the morning to find the rifles gone and his unfortunate comrade lying with his throat cut, as had happened more than once in his experience. The rifles were still there next morning when we went to pull them out, but in a shocking state. They were covered and clogged with wet sand and it was clear that what with the cold, and having the wind up so badly, the man must have been producing bennys at half-hour intervals throughout the night. Our curses were loud and deep as we started cleaning them. But the same thing happened nearly every night.

When the locks and chains did finally arrive we took the bolts out of our rifles and put them in the kit-bags, which we always used as pillows; the rifles were securely lashed with chains to one of the tent-poles and locked with a patent lock. Whenever we were on the march and a man was taken short he handed over his rifle to his section, who carried it between them, until he rejoined them. When on the line of march the locks and chains went with us, but if we did not pitch camp we were ordered to pull our bolts out and sleep with one leg through the sling of the rifle – the very method that had been advocated by Carr since the day the Battalion arrived at Meerut. In most of the camp-bungalows I was in during my service in India strong heavy rifle-racks were fixed, weighing five or six hundredweight and holding between sixteen and twenty rifles. The racks had very heavy lids which at night were dropped and locked over the muzzles of the rifles; strong thin chains were also passed through the trigger-guards and locked. Bolts and bayonets were locked up in the boxes of the men. These precautions were absolutely necessary, for in many stations the bungalows were on the fringe of the jungle or open country, which gave the rifle-thieves every advantage.

In the winter of 1904-5, the First South Wales Borderers, stationed in bungalows at Meean Meer near Lahore, lost a number of theirs and I don't believe that they ever recovered them. This was how it happened. The quarter-guard was in a guardroom with the back-door facing the open jungle. Just before dawn the two sentries, one of whom patrolled behind, and the other in front, had very nearly completed their two hours' sentry-go. One of them came to a stop outside the front-door, and the other outside the back-door, waiting for the next relief. The distant chimes of a clock told them that their time was up. They waited a minute or two and not hearing any move inside one of them opened the door, thrust his head in and shouted "Next relief!"

Sentries did two hours on and four off during the night. Those that were off got down to it for a little shut-eye, but the Sergeant of the Guard and the Corporal under him kept awake in their turns to rouse the relieving sentries. On this occasion everyone in the guard-room was fast asleep until the shout of the sentry woke them, and when the Sergeant and relieving sentries got up in a hurry they had the shock of their lives. They saw by the light of the hurricane lamp which was still burning that there wasn't a Service or Snider rifle in the tent. As the Service rifles were not in use during the night they had been secured in the rifle-rack with a chain running through the trigger guards; the bolts had not been removed, as a guard-room would be the last place in the world that loose-wallahs could be expected to raid. The lid of the rifle-rack was now seen to be raised, and at the foot of the rack, which had contained eight rifles, lay the padlocks of the lid and also of the chain. (The Sniders had not been secured, in order to have them ready for immediate action, if necessary.) And yet in spite of the noise that even the most expert Pathan must have made in picking the locks and removing a whole set of rifles from the rack, not one of the guard had heard a sound. They all complained of having a nasty headache and feeling sleepy. (With the exception of the cries of wild animals the two sentries had also not heard anything.) The loose-wallahs must have sent them to sleep in some mysterious way and got off with the rifles while the sentries were away at the other end of their beats. The Sergeant, who was a very conscientious soldier, was court-martialled and reduced to the ranks, but a good deal of sympathy was felt for him and, as he was one of those who intended to make the Army their profession, he got his stripes back within the space of two years.

ON THE LINE OF MARCH

Lord Curzon came in for strong criticism from all sides. Early in 1903 my company was stationed at Delhi Fort during the great Delhi Durbar held in honour of King Edward VII's Coronation of the previous year. I cannot be sure whether it was the Duke of Cambridge or the Duke of Connaught that represented their brother the King, but whichever of the two it was, the story went around that after the native Princes had paid homage to him he remarked pointedly that from their behaviour it seemed as if somebody had been spoiling them: they did not behave in the same manner as during his last visit to India.

A great parade was held on the maidan outside Delhi. Sixty or seventy thousand British and native troops marched past. Special medals were issued on this occasion but the men did not think much of them, any more than they thought much of the Good Conduct medal, or "Rooty Gong." (This was awarded, as the saying was, for twenty years of undetected crime and had a ribbon attached that too closely resembled that of the Victoria Cross. "Rooty" is Hindoostani for bread, and the medal was so called because it was a regular ration-issue, like bread or meat or boots.) Among the troops who marched past was a certain Lancer regiment which not long before had been stationed at Peshawar. One of their men had been murdered in the native bazaar and the following evening about a dozen of his comrades decided to take the law into their own hands and avenge his death. They saddled their horses and with levelled lances charged through the bazaar where the murder had taken place, and, avoiding women and children, stuck as many natives they met as were not quick enough to dart away behind the booths. They stuck quite a few. I don't know what happened to this party of men, but the regiment as a whole was reprimanded by Lord Curzon, who ordered their removal to a punishment station: a dreary spot somewhere, hot, and none too healthy. It is said that when the order "Eyes right" was now given as the regiment passed Lord Curzon at the saluting base not many men obeyed the order, and some even reversed it.

One incident that I remember of this Durbar was that a magnificently built half-caste prostitute of fifty years of age chose the date to announce her forthcoming retirement from business. She had been a prostitute for thirty-six years, most of them in the brothel reserved for the white troops, and the life had suited her well. She had saved enough of money now to be able to call her time her own. To celebrate this happy day and also out of loyalty to the Crown she decided to make a positively final appearance that night and give all soldiers who wished to take advantage of her offer free access to her body between the hours of 6 p.m. and 11 p.m. Preference was given to old customers. She posted notice to this effect on the door of her room and if I related here how many men applied and were admitted and went away satisfied in those short hours, I should not be believed. I remember reading in Gibbon's *Decline and Fall of the Roman Empire* about the Roman Empress Messalina who was also a star-turn in the brothels of Rome; but if that half-caste woman had been living in Messalina's day she would doubtless have fallen a victim to the Empress's jealous rage.

The native Princes showed their loyalty not only by their homage but by taking part in the Elephant Procession (which I did not witness), each Prince doing his best to display as much wealth on his person as he possibly could. Durbars come very expensive to native princes and they are therefore quite sincere in wishing the reigning sovereign long life; and their subjects, who are more heavily taxed in Durbar years than usually, also offer the same heartfelt prayers.

I had not seen rain since I had been in the country, but early one afternoon I thought I was going to see a rainstorm that would beat any I had seen in my life. In the distance a wall of blackness reaching from the ground to high up in the sky was gradually approaching us. I said to Carr, "This rain is going to wash the bloody camp away." Carr said, "That's not rain, that's locusts. And they're liable to do a damned sight more damage to the country than what they would if they were rain." Fortunately the storm, which lasted for hours, swept right over us without settling. North, south, east and west was a wall of blackness. Locusts are something similar to grasshoppers, only very much larger. After the sky was clear again there were thousands of them perched on the tents. Probably these were the weak ones who had fallen out on the line of flight; because they made no effort to fly away when we started slashing them off the tents with our towels. Carr said that it was the worst locust storm he had seen, and that if they had settled they would have snowed us under. I never afterwards saw a worse one myself.

About the middle of March we made preparations for leaving Meerut to march to Chakrata, a hill-station in the Himalayas one hundred and sixty miles away. About a fortnight before we began this march very heavy rain fell which lasted some hours, and the camp was soon flooded out. After the rain ceased the camp was invaded by hundreds of water-snakes, some of them nine feet in length. They were not dangerous, but they gave me the creeps. When the water subsided they vanished back into the holes in the ground from which they had come and we did not see one of them again until about the same time the following year, when we again had a heavy fall of rain. We struck camp. The large marquees were removed and we were issued with mountain-tents which could be erected in five minutes and struck and packed up in the same time. Large bags called "sleetahs" which held the kit and blankets of four men were issued out: these were carried by the Battalion transport on the line of march. The heavy baggage was dispatched by rail to Dehra Dun, the rail-head for Chakrata. Transport in India was supplied by the Indian Supply and

Transport Corps, which was either bullocks, camels or mules. Bullock transport, which we had on this march, was very slow. We always had to wait an hour or two for the wagons after we had completed our day's march. Camels were much quicker, but the mules were quicker still and arrived in camp on the heels of the Battalion.

The dairy, bakery, cooks and camp-followers moved off each evening twelve hours in advance of the Battalion, so that rations could be drawn and breakfast ready by the time the Battalion arrived. There was no breakfast before we started out on our march, which on some days was stiffer than on others, but any man who chose to do so could give his name to the Colour-Sergeant who would put it down on the list of men who would be daily supplied with a good meat-sandwich and a pint of tea at the coffee-halt, for which two annas a day was deducted from their pay. The Battalion coffee-bar supplied the sandwiches and tea, which were issued out after half the day's march had been completed. We always knew when we were approaching the coffee-halt, where we had half an hour's rest, by the drums striking up with the tune of "Polly put the kettle on and have a cup of tea." There did not seem much difference between the line of march and a standing camp; the mucking and other wallahs came around shouting their eatables, the Canteen was opened at the usual time and the shopkeepers from the Regimental Bazaar, who also travelled in advance, had erected a smaller edition of their shops. It was rarely that we pitched camp; the nights were warm enough to sleep out in the open.

We started each day's march at dawn and the only parade we did after arriving at camp was rifle-and-foot inspection. Unless a man was on guard he had the rest of the day to spend how he liked. We were allowed in the small villages near which we camped, but a large-sized town was always out of bounds for us. Most of the men passed the day away by playing House, which was the most popular of the games played. The Prayer-wallah and I preferred it to any other: it was the most sociable and one could not gain or lose much at it, and it was leisurely and long drawn-out. Crown and Anchor, Under and Over, Kitty-nap and Brag schools were also dotted over the camp. The majority of the men were inveterate gamblers and those who were stony broke would collect in schools of five and play Kitty-nap "for noses," as it was called. When a player called Nap and made it he would bunch his five cards together and give each of the others twelve smacks on the nose with them; if he failed to make his contract he received six smacks on his own nose from each of the others. Everyone put the full weight of his arm behind each smack, and a man who had a bad run of luck with his calls and was blessed with a large nose would receive severe punishment. His nose would be smacked downwards, upwards and sideways until the water would dance out of his eyes.

There are many different castes, languages and religions in India. The Mohammedans bury their dead; the Parsees put theirs on a hill for the birds to pick – a clean method because the vultures and other large birds strip the corpse of its flesh in a very short space of time; the Hindoos burn their dead until only the ashes remain. After one day's march the Prayer-Wallah and I took a walk over some open country which brought us not far from a primitive native village. About a mile from the village a group of natives was clustered weeping and wailing around a raised iron bier under which a fire had just been lighted. We watched the scene from behind a large tree about fifty yards away. As the fire became fiercer, the corpse of an old man which was on the bier rose slightly up. He was at once jabbed down again by a native who was holding an implement like a large hay-fork. This caused the Prayer-wallah to say, "Stone me pink, Dick! They're burning that poor old bastard before he's dead!" If I had not seen a similar thing before I should have agreed with him; but when on the way up from Deolalie a few of us going for a walk one day had seen a corpse, who was being burned, behave in just the same way. One man, who seemed to know something about it, had then explained to us that the heat of the fire caused the gases to operate in the corpse, which made it rise slightly. I told the Prayer-wallah this, but he had his doubts about it. He said that he had heard that, once the funeral prayers had been said over a corpse, it was officially dead, and that if a mistake had been made, the person could not be received back into the family as a living person; so unless the corpse was very much alive and jumped off the bier, they found it best to make sure of it by jabbing it down with the hay-fork which they carried for this purpose. We got to arguing about the behaviour of chestnuts in the fire and sausages in the pan; but the Prayer-wallah said that Nature had left holes in a man on purpose for the gases to escape, and that if you nicked a chestnut or a sausage with a fork it would stay put, because the gases had an outlet.

At every village where we stopped, the whole of the villagers would turn out, but they paid more attention to our Regimental Goat than what they did to us. They believed that he was one of the Gods whom the Battalion worshipped. The Goat travelled with the transport, riding all the way in a small bullock-cart drawn by a handsome white bullock. On arrival at camp he was trotted around for a little exercise before he was given his breakfast, after which he was washed and combed. During these operations the villagers looked upon him with awe. No wonder they believed he was one of our Gods, with all the care and attention that was bestowed upon him. Our Goat had been about six years with the Battalion at this time and the King's herd of big white Kashmir goats at Windsor, where he had come from, had never bred a more handsome, wicked and lazy goat than what he was. This herd had been a present to Queen Victoria, on her accession, from the Shah of Persia. A man called the Goat-Major, a lance-corporal, looked after him. The Goat attended all ceremonial parades, and marched to Church and back in front of the Battalion with the Goat-Major leading him. About a mile was as far as he would march and all the goat-majors that were ever born would never have made him march further.

The natives attached to the Battalion were in mortal dread of the Goat. The vegetable-wallahs, who came around camp at Meerut carrying baskets of vegetables on their heads, kept a wary eye open for him as they walked up and down the lines during the afternoon, and prayed to their gods that he was securely tied up. There was hardly one of them that he had not knocked half-unconscious at some time or other; by the time they had recovered their wits there weren't many of their vegetables left. He had a little mountain-tent of his own, not far from the quarter-guard. During the afternoon he was tied up outside the tent with a rope which was long enough to enable him to walk in and out of it. On a number of occasions he either slipped or broke his rope early in the afternoon; after doing this he would stroll unconcernedly down the lines, dodging in and around the tents until one of the vegetable-wallahs was within charging distance. This Goat had sense and craft in abundance and took advantage of cover like a Pathan. When his victim was about twelve feet from him he would manoeuvre around a tent so that he could deliver his charge from the rear. It was rarely that he failed to connect with his victim's backside and send the basket of vegetables flying. After this had happened (but never before) the Goat-Major Would come running down the lines shouting, "That blasted goat has broken his rope again!" I always believed that Goat and Goat-Major had a perfect understanding, because if there were no senior N.C.O.s about the Goat-Major would collect most of the lettuces and gibbons that were strewn about; these provided a nice cheap tea for himself and luxuries for the Goat.

I don't know whether the vegetable-wallahs were recompensed in any way for their loss. If so, they were lucky. The Goat-Major, who was not expected to be with the Goat all the time, said that if the Goat were handcuffed and manacled and tied in a sack he would free himself as easily as Houdini, once he heard a vegetable-wallah crying his wares in the camp. With a single exception, not one of the vegetable-wallahs were seriously hurt by the Goat's charge, and even this exception was a case of physical injuries indirectly inflicted. He was a frail Mohammedan, who had a small harem of six wives. How he managed to provide for them was a mystery: he seemed to make hardly enough of money to provide for himself. The other vegetable-wallahs looked upon him with scorn because according to bazaar gossip his wives sometimes combined to give him a beating. The Goat only had occasion to charge him once, but when this happened he pitched such a tale of woe to his wives when he arrived home that they rose up and gave him such a beating that the frail Mohammedan did not appear in the camp for several days. When he did appear he was in a very serious state. If half a dozen cheetahs had clawed his face they could not have done it to better order than his six wives had done.

A number of girls from the brothel in the Suddar Bazaar at Meerut followed the Battalion on the march. They made the journey in all sorts of conveyances and a man did not need to walk far from camp before being accosted by one of them. No doubt the money they earned more than covered the expenses of their journey to the Hills. Many of them spent the summer there regularly: it was hot, carrying on their profession down in the Plains. Indian fakirs sometimes appeared in the camps. They did a number of extremely clever tricks in the open air and the collection of two or three rupees which they always made up before starting their performance quite satisfied them. During my service abroad I saw many fakirs but not one of them could do the celebrated rope-trick. If this trick can be performed I was unlucky enough never to have witnessed it. The old Bacon-wallah told me that he had only seen it performed once and that was during the early years of his soldiering in India. According to his story, his Battalion was visited at their camp one day by an elderly fakir who gave a performance in front of the large Officers' Mess tent. Many of the men, including himself, witnessed it. After he had concluded his performance one of the officers asked him if he could do the rope-trick. He replied that he could, but an extra collection of thirty rupees must first be handed to him.

When he had counted out thirty rupees he produced a coil of rope from his basket and uncoiling part of it threw it in the air. The coils straightened out and the rope, which was about thirty or forty feet in length, hung in the air with one end of it about two feet from the ground. The fakir now told a small native boy who was with him to climb the rope. After he had climbed to the top of it he pulled the remainder of the rope up to him. The fakir made a few passes with his hands and boy and rope vanished in the air. Then he told the onlookers to look towards some bushes about fifty yards away, out of which sprung the native boy who had climbed the rope. He quickly ran to the fakir, but he had no rope with him. There were many natives attached to the Battalion who were spectators of the performance. They said that fakirs of his class were only seen once during a man's lifetime and that the men who had witnessed his performance would never see him again, once he had left the camp.

Carr, who told a similar story, said that he, too, had seen the trick, but only once. And when we were at Agra one of our chaps who was a bit of magician himself went on leave to Benares, and on his return, said that he had seen the rope-trick performed there by a fakir who was the King-pin of all magicians in this world. This man was not given to romancing. I had often heard him say before he went to Benares that he would gladly give a year of his life to see the rope-trick performed. There are many explanations of the rope-trick. Some say that it is done by hypnotism and that once somebody took a camera-snap of the fakir performing the trick, and neither boy nor rope appeared on the finished photograph – only the fakir himself gesticulating and the audience with a glassy look in its eyes. Others say that the coil of rope is really made of hundreds of small pieces of jointed bamboo, painted to look like a rope, and that when the fakir throws it up in the air he pulls a cord that runs through it, which locks the rope into a rod which a boy can climb up. They say that the trick is always performed under a high tree and that the boy disappears into the foliage drawn up by a confederate on a thin piece of wire; and pulls the rope up after him, while he is rising on the wire: the fakir has unlocked the joints so that it can be rolled up again like a rope. But it would never have done to tell the Bacon-wallah or Carr or the man who went to Benares that it was only a simple conjuring trick: they liked to feel that they had witnessed a piece of genuine Eastern magic.

One day a troupe of tight-rope and slack-wire walkers appeared in the camp. The outstanding performer was a youth of seventeen who did stunts on the slack wire that would have made all the professional slack-wire walkers in Europe turn green with envy. For an extra collection he produced two hollow pieces of horn which had flattened points on them about the size of a threepenny-bit; they were just large enough to cover tightly the toes and the broad pan of the feet. He fixed these on before mounting the platform, which was about fifteen feet high. He then slid along the wire and did tricks on the tips of those horns as easily as he had done them on the flat of his feet. I never saw this troupe again, which would have made pots of money if it had toured England, and among the many others I saw there was no single performer who could walk a slack wire, like this youth, on the points of horns.

We passed through some interesting places before we reached Dehra Dun, which was sixty miles from Chakrata. That afternoon the Prayer-wallah and I met a party of natives carrying a dead tiger which had been shot by a shooting party only two hours before we met them. The tiger was about eight feet in length. They told us that a month previously one of the largest tigers that was ever shot in India was shot just outside this place – it measured twelve feet nine in length.

Four companies of Ghurka soldiers were encamped at Dehra Dun – a short sturdy race of men closely resembling the Japanese; they were proud of the fact that they were the highest paid of all Native Infantry. In addition to a rifle and bayonet they were also armed with a large curved knife called a Kukri, which many of them could throw accurately at a given target. The Kukri was a symbol of honour for them, and they attached much the same importance to it as what I have read the Greeks of old attached to their shields – they would rather die than lose it in battle. They came from the mountainous country of Nepal, the most truly independent kingdom in India. As a result of a treaty that Lord Curzon made with the King of Nepal, no white man is allowed to settle there except at the King's personal invitation. For every Ghurka who enlists in the Indian Army the Nepalese Government receives a bounty of two or three pounds. I don't remember ever hearing of a Ghurka battalion being stationed in the Plains, which were too hot for them. They lost more men from pneumonia than any other disease; the hotter the station they were in during the summer, the more men they lost from this. In the summer it was hot enough at Dehra Dun, which was called a second-class Hill-station, but not so hot by ten or twelve degrees as some of the Plain-stations. The Ghurkas were merry little chaps and the only native troops with whom British soldiers were friendly enough for joking and playing tricks. They were proud of talking a little English and it is curious to think that in Nepal, where no Englishman can go, English is spoken as a second language in the Ghurka villages by the families of old soldiers.

It was four days' march from Debra Dun to Chakrata and on the second day we arrived at a delightful village called Kalsi built in a clearing of a forest at the foot of the hills. Kalsi was quite free from that peculiar sweaty, spicy smell which all cities, villages and bazaars have in the East. Around it were tea-plantations which provided work for the villagers. There were some lovely streams running through the forests and abundance of game. Not far away the river Jumna rises. On the banks of one of these streams, with a little wall surrounding it, was a huge stone embedded in the ground, that must have weighed several tons. Cut all over it were strange hieroglyphics which the learned experts had so far failed to decipher. The villagers' story of the stone was, that over five thousand years ago a mighty King of Northern India visited the spot and was so enraptured with it that he built a large city around it for his capital and removed his court there. Some years before he died he decided to bury the greater part of his treasures. He ordered a great square pit to be dug, and lined it inside with stone walls. He then had four huge stables built against each side of the pit at the bottom, huge enough to hold four of the largest and fiercest elephants that he possessed. When everything was completed the elephants were lowered into their stables with enough of food and water to last them until the end of the world. The treasures were heaped in the middle of the pit which was then roofed over with a stone vault, upon which the earth was heaped. The elephants remained there on guard. The King was not yet satisfied with the precautions he had taken to safeguard his treasures, so he had this great stone brought from over a thousand miles away to seal the top of the vault. Twelve months later the King died, and a few hours after his burial a great earthquake destroyed the city and every living thing in it. The only object that escaped for miles around was the great stone. The simple natives believed that these four elephants were still alive and kicking, and ready to destroy anybody who attempted to remove the stone in order to get at the treasures.

There were thousands of wild monkeys around Kalsi; they were not much larger than the monkeys that are carried by some of the Italian organ-grinders at home. It was also a bad place for snakes, and notices were put on the trees warning men to beware of them. In Bengal alone thousands of natives died every year of snake-bite, but I never heard talk of any British soldier who had died in this way. During the whole of my service with the Battalion we never had a single man bitten by a snake of a poisonous kind. I could never understand why so many natives died of snake-bite, while British soldiers escaped scot-free, until I asked an old snake-charmer. He explained that snakes have a certain sense of hearing but are all slightly deaf. When soldiers are marching or travelling over open country or through jungle the noise that they make by wearing heavy boots can be heard by snakes, which quickly scuttle out of their way. With natives it is different: they wear no boots and hardly make any noise when working in fields or travelling over open country or through jungle grass: snakes, not hearing their approach, are liable to be trodden upon and, thinking they are being attacked, immediately retaliate with their poisonous fangs. Cobras, which are about five feet in length, are very deadly snakes. The largest I have seen was killed by two of our chaps one morning at Meerut about twenty yards from the latrines. It measured exactly six feet in length.

We did not pitch our tents at Kalsi but slept in the open, the night was warm enough for us not to need more than a single blanket over us. The man who was lying to the right of me had snakes on the brain and kept a few of us awake for quite a time talking about the danger of sleeping in the open in country like this. He only ceased after one of the men fiercely threatened to close his trap if he did not close it himself. A little later most of the men seemed to be asleep, even the man on my right, but he was very restless and wriggling to and fro, as though he was dreaming of snakes. The stillness of the night was occasionally broken by the sharp challenge of a sentry or the cry of some night bird. There was one peculiar cry I had not heard before: it disturbed me once or twice just as I was dropping off. Then the cry of a cheetah sounded quite near: the cheetah could not have been far off, by the way a dog near me was trying to work himself under his master's blanket. It is a queer experience to lie in the midst of a group of sleeping men and be unable to sleep oneself. Some were groaning or talking or making queer noises. One man began chuckling and laughing so heartily that I felt like laughing myself. But another of them commenced to grind his teeth and I felt like murdering him. I don't think there is any sound in Nature more horrible than a sleeping man grinding his teeth – unless it is a sleeping woman when she grinds hers.

I thought that I was the only man awake and began wondering if I would ever drop off. The man grinding his teeth ground them more desperately than ever, as if he was under torture or trying to chew a particularly tough khaki patch, and the Prayer-wallah, who was lying with his head not far from my feet, began to pray in a low but distinct voice. He earnestly requested the Almighty of His Infinite Goodness and Mercy to send down from on high a first-class lightning-dentist with strict orders not to report back at the Orderly Room of Heaven until he had painfully extracted every blasted tooth, sound and rotten, upper and lower, single and double, from the grinding-teeth-wallah's head, so as to give His faithful flock the blessings of deep and peaceful sleep so long despaired of.

Half a dozen of us said, "Amen, Amen!"

"Christ, I thought I was the only poor unfortunate swaddy awake," he replied. Each one of us had thought the same until he commenced to pray.

We laughed and that set off the laughing-wallah again. He had such a hearty fit of laughing in his sleep that it caused most of the men around him to wake up, including the senior N.C.O. and the man who had been grinding his teeth.

The teeth-wallah, as he was called now, began to curse and said that he had not slept a damned wink during the whole of the night; it was not the first night that the bloody laughing hyena had kept him awake and if he had his way he would gag his mouth with a rolled pair of bloody socks every night before he got under the blankets.

This was too much for the Prayer-wallah, who told him that he would much prefer listening to a whole pack of laughing hyenas than a man who ground his teeth like a thirty-foot shark just waiting for a ship to go down – like what he had been doing before the laughing-wallah woke him up.

A battle of words ensued, together with much bad language in several dialects. We others who had been kept awake by the grinding of teeth also joined in the battle on the Prayer-wallah's side, each one telling the teeth-wallah what we would like to do with him. He was not a physically strong man, but he was passionate. He threatened to bayonet anyone of us who attempted to get rough with him. Quiet was not restored until the N.C.O. had threatened to make us all prisoners if there was not instant silence. The laughing-wallah had not been disturbed at all and went on chuckling and laughing in his sleep, not knowing the part he had played in this row.

After twisting and turning for some time I finally dropped off to sleep. I could not have been asleep very long before I, and half the Battalion with me, were awakened by the man on my right screaming, at the top of his voice, that a snake had crawled under his blankets and bitten him on the knee. He seemed absolutely terror-stricken and we had a job to calm him. We hurriedly lighted a candle and examined his knee, but found no trace of a bite. He then looked a bit foolish and said it must only have been a dream, but it was so vivid that he could hardly believe it wasn't true. He had put the wind up a number of men who were now on their feet and beating and shaking their blankets for all they were worth. The row they made would have frightened away all the snakes for miles around. Afterwards we reckoned that the dream had been caused by the man sleeping with his leg through the sling of his rifle: the sling was the snake and as he turned in his sleep his knee had come in contact with the backsight of the rifle which gave him a slick prick and caused the bite. Some of the men were most unsympathetic, especially the man who had threatened earlier on to close his trap; he now told him that if ever he happened to notice him in the grip of a gigantic python who was slowly crushing him to death in his coils before swallowing him, he would much prefer that sight to seeing all the treasure that the natives talked about in connexion with the huge stone laid out shining at his feet. The Prayer-wallah addressing no one in particular, said that the screams of the snake-wallah were heavenly music compared to the sound of the teeth-wallah had been making before he was awakened by the laughing-wallah. This caused another laugh and also another argument. But the argument did not last long: the N.C.O. now said that if he heard only a whisper from a man he would run him straight to the Guard-room. By the time I had just dozed off again, reveille blew. Considering what a night's rest we had had, it was surprising to find us all in such good spirits to begin the day's march.

There were two routes from here to Chakrata. One, called the short cut, was a stiff uphill climb of about seven miles followed by a slight descent to a place called Chilmeri Nek which was about a mile and a half from Chakrata. Only men and pack-mules could travel up this track, which we took the following year. The other route was a good road cut in the hillsides, which are called khuds; it has a gradual rise and winds like a snake for sixteen miles before it reaches Chakrata. After we had marched about five miles we saw a large bungalow on a hill not far off: one of our officers said that it was the hospital for the troops at Kilana, which was only a mile from Chakrata. At this point the road took a sharp bend and we did not see the hospital again until we had marched about another five miles. It then looked just as far away as it had on the first occasion we saw it and more marching brought it no nearer. By the time we finally reached Chakrata we were all agreed that it was the longest sixteen miles we had ever marched. The Prayer-wallah said that the man who had measured the road had ridden downhill all the way mounted on a Derby winner and after galloping each three miles had put down one for it.

A HILL-STATION

The bungalows at Chakrata were scattered here and there on the hillside, the highest being about eight thousand feet above sea-level. There were two squares, called the Top and the Bottom Squares. On the Bottom Square, where the whole of the Battalion sometimes paraded, cricket, hockey and football were played. At Chakrata, it did not matter where a man had to go, he would always have to walk either uphill or downhill. On a hill between Chakrata and Kilana stood the Protestant church for both stations. To the north of us stretched a magnificent range of mountains covered with perpetual snow and extending for hundreds of miles. Mighty peaks, ranging from eighteen thousand to over twenty-eight thousand feet above sea-level, were in this snowy range. Some days the range did not look thirty miles away, yet the distance was one hundred and fifty miles. From one or two of the higher hills in the neighbourhood Mount Everest could occasionally be seen: on very fine days it looked like a misty dome behind the snow-clad peaks, to the north-east. Everything was so vast and the atmosphere was so clear that for the first three months at musketry-instruction we always under-estimated the distance to a given object. Some of the khud-sides were bare and stony, but most of them were well wooded. Often they had a steep and dangerous descent of thousands of feet into the valleys below, which were full of wild fruit-trees of all descriptions. In the wooded khuds lived large tribes of monkeys, who were about four feet in height. Some of them, whom we called the grandfathers of the tribe, had long white beards: looking at them you felt that they ought to go about more decently dressed.

On one steep, inaccessible cliff hundreds of eagles nested; even the smallest of the full-grown ones had a seven-foot span of wing. I saw one that had been shot, with a ten-foot span. They were silver-bellied eagles, and they always interested me when they were flying and screaming overhead. Whenever they turned over in their flight their bellies were like flashes of silver in the rays of the sun. One Sunday afternoon the Prayer-wallah and I set out on a long tramp together, with our khud-sticks in our hands. These were bamboo poles about five feet in length with sharp iron spear-points fixed on the bottom of them: they greatly assisted a man when khud-climbing. After tramping for some hours, we were making our way along a narrow track on the edge of a steep bare khud when we heard a violent screeching from below us, mixed with dull heavy thuds like the noise of carpets being beaten. Wondering what all the row was about we got down on our stomachs and peered over the edge of the track. About two hundred feet below us, on a very large flat rock jutting out of the khud-side, were two full-grown eagles engaged in a battle royal; by the look of them they had only just begun it. They could not have picked a better spot: with the exception of ourselves, the only living things in sight were a number of other eagles flying high up about five miles away. Although the khud was steep and dangerous to descend we could have got closer to them, but we were afraid of dislodging stones which might have disturbed them, so we remained where we were. They darted in and out at each other, using their beaks and talons to some order, and the noise they made when they were buffeting each other with their huge wings could have been heard a great distance away. After they had been interlocked for a while and broken loose again we could not tell which was which, or whether one or other was getting the best of it; they were of equal size and seemed well matched. After the fight had been in progress for quite a time one of them, who seemed to be weakening, suddenly rose into the sky. He had evidently had enough of it. The other went swiftly in pursuit and soon caught him up, flying a little above him; as he did so he made a fierce downward dart at the neck of the beaten enemy and then the both of them seemed to drop like stones into a valley out of sight from us. I expect the victor saw the beaten one safely to his death at the bottom; as afterwards in the War I once saw one of our planes diving after an enemy plane which had caught fire, making sure that it crashed properly.

We had been so interested that hardly a word had been spoken between us while the fight was in progress. The Prayer-wallah now said: "Well, Dick, it's worth doing eight years in the Army if it's only to see a fight like that! I wouldn't have missed it for a thousand rupees." I agreed, but I had felt all along that the sight of the Snowy Range was enough by itself to make me satisfied that I had enlisted. I now proposed that we should make our way down the khud for a closer view of the battleground. We started down at once, but the slope was so steep just below this point that we had to make a wide detour before we arrived at the rock. When we reached it we both gazed in wonder at the great splashes of blood all over it, and asked each other how the birds could have flown at all after losing so much blood: what struck us also was the small quantity of feathers that they had lost, considering how viciously they had used their wings. I have seen many fights in my life, between men, and between animals, and between birds; but none ever to equal that one.

Black bears roamed the hills and natives sometimes came around the bungalows with puppy bears for sale. Several were bought by our men at ten rupees each, but when they grew to their full size they became dangerous and had to be destroyed. Quite a number of men possessed dogs which they had bought at Meerut; there were half a dozen breeds in some of them and one, that was a pure cross between a bulldog and a greyhound, looked a fearful creature. A man in my tent at Meerut had bought a very clever little monkey and dressed him up with little striped trousers, red coat and a pillbox on the side of his head. He gave him a little wooden musket too and trained him at the word of command to go through all the arms-drill that a soldier was taught. He had a small collar around his neck, to which was attached a long thin chain. During the day he was tied up with this chain to a large tent-peg outside the tent; on cold nights he slept at the foot of his master's bed. The man badly wanted to see what effect a drop of beer would have on his pet, so one day he brought about a pint and a half of beer in a basin from the Canteen and held it for him to have a drink. The monkey took a good drink and the way he smacked his lips afterwards made some of us who were looking on think that it was not the first occasion that he had tasted beer. By the time he had drained the basin dry he was helplessly drunk. He staggered towards the tent-peg to lean his arms on it, which was his usual custom when resting during the day; but he must have been seeing a dozen pegs, because each time that he put out his arms to lean on it he was still two or three feet away. After falling down half a dozen times, he gave it up and the last time he fell he went to sleep. He now took the habit of accompanying his master to the Canteen every evening; after he had performed a few tricks he would go along from table to table, holding out a little tin mug for a drop of beer to be put in it. Night after night he got gloriously drunk, and after he had been with us twelve months his master awoke one morning to find him dead at the foot of his bed. All the boozers were convinced that he had drunk himself to death, which in their opinion was the most noble and happy end to which either man or monkey could come.

During the years I was with the Battalion there were some well-known boozing schools. A boozing school generally consisted of three or four men who pooled their pay, one of them acting as treasurer. They allowed themselves so much for tobacco or cigarettes and so much for a monthly visit to the women in the Bazaar; the remainder was spent on beer. Only one basin was used between a school; it held a quart and each man took a drink in his turn from it, and each in turn walked to the bar with it when it wanted refilling. When money ran short they would borrow money right and left and sell any kit they did not want and also some that they did, so that they could be present at the

Canteen when it opened. Genuine boozing schools always paid their debts. They would borrow to the extent of two hundred rupees, which was their limit: once they were that amount in debt they turned teetotallers, or "went on the tact," as it was called. During the time they were on the tact they still pooled their pay and took long walks together. The money set aside for tobacco and women remained the same but they lived simply on their rations. After they had paid their debts they would decide among themselves whether they would carry on with the boozing school or not. If they decided to carry on with it they would still continue on the tact until each of them had bought enough of clothes and kit to last him for years. (A best suit of khaki or white only cost six rupees.) After being on the tact for about six months they would start boozing again with a capital of two hundred rupees: this was the savings that remained after buying the kit and clothes – which they would sell again as soon as they were hard up. Some schools broke up after paying their debts, especially if the members had only twelve months to serve before becoming time-expired. During that twelve months they usually "went on the skin," as it was called, which meant that hardly a pice was spent on anything. Two men, whom I knew very well, even denied themselves a smoke in their endeavor to save as much money as they possibly could before returning to civil life. These two were expert knitters: they bought wool from the Bazaar with which they knitted jerseys with fancy designs, and blow-belts, for keeping money in, which a large number of men wore around their waists. They made a handsome profit out of their work and considering how much they had liked their beer and tobacco it was marvellous how they stuck it so well on the skin. Yet they were only saving their money, so they said, to have the booze of their lives when they landed in Blighty.

The Battalion did a lot of khud-climbing, which kept the men in good condition. If a man had had a good skinful of beer overnight, a few hours of khud-climbing would soon bring it out of him. We signallers did our own work, which I found very interesting, especially when we were working with a five-inch helio or lamp. A helio in these hills could easily be read at a distance of sixty miles with the aid of a telescope. The lamp, which burned oil, was called the "B.B." and could be read at a distance of ten miles with the naked eye and twenty with the telescope. Thirty-two miles from Chakrata, as the crow flies, were two Hill-stations adjoining one another, named Landaur and Mussoori. On a clear day we could make out the bungalows at these places. We worked in with the helio station at Landaur and sometimes, for practice, the signallers of the units stationed at Kilana would form a transmitting station between us. On a clear day we never used the telescope to receive a message; but a slight mist made it necessary. The rarity of the atmosphere had an effect on the hearts of some of the men. We had not been there a month before forty of them, after being examined by the Medical Officer, were sent back to Meerut. He said that there was nothing seriously wrong, but that the extra deep breathing that they had to do was causing them to overwork their hearts. At Meerut they were attached to the Rifle Brigade, and did not find themselves at all comfortable there. Many of them got into trouble from following the Royal Welch regimental tradition rather than that of their hosts, and a few received sentences up to two months' imprisonment. The following year the same thing happened to their hearts and after a stiff bit of khud-climbing their faces would go all colours. But not one of them reported sick. They would rather have their hearts crack, they told us, than be sent away from home again to live among strangers.

Our bungalows were partitioned off into small rooms, each room accommodating between sixteen and twenty men. There were open fireplaces in each room where, after the arrival of the monsoon, fires were constantly kept burning. A daily ration of wood was drawn by each company at the Ration Stand and distributed to the rooms. The wood was issued in three-foot logs and each room had a decent number of logs stacked outside on the verandah by the time the monsoon arrived. Every morning quite a number of Hill-men would congregate at the Ration Stand soliciting the job of carrying the wood to the bungalows. For two annas they would carry 180 lbs. of wood on their backs up a steep hill to a bungalow a mile away. They always had their own ropes to lash wood or anything else they had to move, and carried everything on their backs in much the same way as a soldier carries his pack.

The natives around this part had a Mongolian cast of features and were more yellow than brown. Most of them had delicately tinted rosy cheeks, which I expect came from the open-air life and the high altitude they lived in. The men were strong and muscular, the majority of the young women very handsome. I have never seen women with a more erect carriage and graceful walk than those from the villages around Chakrata. They were a peaceful people and had to work very hard on their plots of land during the summer to enable them to live during the winter, when they were more or less snowed up. Some of them possessed cows, sheep and goats, but many of them had a job to earn a living. I don't remember what religion they followed but they were strong believers in a plurality of wives. Some of them had half a dozen wives; they worked hard themselves and made their wives work hard as well. It was a common sight to see a man with his wives behind him in Indian file walking along the road or along the tracks of the khuds. There was a case of two brothers who were so poor that they could not buy a wife apiece. They got out of their difficulty by pooling their money and buying one between them. No doubt most of the wives of the men who possessed six or more would gladly have exchanged places with her: she was receiving so much and they were receiving so little of one of the main pleasures of life. But, as time went by; the two brothers, with the help of their wife, saved enough of money to buy another wife; which they did. On the day of this new purchase they had decided to start separate households with a wife apiece. They must have been a sporting pair of brothers, for they had a little gamble – the winner to have his choice of the wives, who had no say in the matter at all. The winner chose the fresh young bit. As the first had not had a child matters were simplified; if she had had one I expect the loser would have reared it up, whether he had been its father or not.

Whenever they held a festival the women would dress in their best robes, on which they stitched rupees. The more rupees they had stitched on, the more attractive it was supposed to make them look. I saw one favourite wife of a well to-do man who had three hundred and fifty rupees stitched on her robes, and huge silver bracelets on her arms and ankles. She looked like a walking mint and how she carried so much weight was beyond me. Like all other women in the world she would have cheerfully carried a ton of precious metal if she thought it would make her look more attractive than those of her sex who were carrying only about fifteen hundredweight.

The Prayer-wallah and I decided to go on the tact for a month. Our main object was to save the greater portion of our pay so that we could buy an Indian tablecloth and a couple of Kashmir shawls apiece to send to our folk at home. I spent most of the time that I would otherwise have passed at the Canteen, in reading whatever books I could get hold of. One of the few things I had against the Prayer-wallah was his taste in books. Historical romances were what most appealed to me. There are few of Dumas's novels that I have not read once or twice: Dumas was always my favourite. Stanley Weyman's *A Gentleman of France* and *Under the Red Robe* I also liked. But the Prayer-wallah swore by Hall Caine, which was something that I could never understand: Hall Caine seemed to me more a novelist for ignorant servant-girls than for soldiers who had knocked about as much as the Prayer-wallah and I had done. Dickens' books were easy to get hold of and I read many of them, but except for *Pickwick Papers* and *A Tale of Two Cities*, both of which were out of his ordinary line, and the first part of *David Copperfield*, they disappointed me. In Dumas's *Three Musketeers* or *The Count of Monte Cristo* there is no waste matter and the story is exciting, and you do not keep forgetting who is who, as in some of Dickens' works. The Prayer-wallah spent his time in learning the "crab-bat" as it was called, which was all the swear-words in the Hindoostani language and a few more from the other Indian dialects to help these out. He had picked up a fair knowledge of the crab-bat at Meerut but he now studied it seriously and used to curse the natives, whenever they deserved it, to such order that they looked upon him with veneration and praised him as the oldest of old soldiers.

It must not be thought that the Prayer-wallah and I were exceptional in the matter of reading books. There was a good lot of reading done, especially of the racing tales of Nat Gould. Nat Gould had written at least fifty of these paperbacked novelettes, and when a man had read one he had read the lot; but the titles and the names of the characters were always different and the hero's horse, which got nobbled yet always managed unexpectedly to pull off the big race at long odds and thus save the heroine from the clutches of the villain, was sometimes a handsome chestnut and sometimes of another colour altogether; and this made for variety. Then there were the novelettes of Paul de Kock which were addressed to the baser passions, as the parsons would say, and which, like Nat Gould's works, one could always buy in the Suddar Bazaar: the native shopkeepers did a roaring trade with these and also with a cheap version of the *Droll Stories* of Balzac. As for the *Decameron* of Boccaccio, in my time every soldier of the British Forces in India who could read had read this volume from cover to cover. It was considered very hot stuff; but the Prayer-wallah used to say that in this respect it did not come within shouting distance of certain passages in the Old Testament, once you got the hang of the Biblical language. Kipling's works were also to be had from the Library. What interested us most in them was his detailed accounts of native life; his knowledge of barrack-room life seemed to us a bit hazy in places, but perhaps that could be put down to *Soldiers Three* having been published many years before, so that times had changed in the meanwhile. At any rate *Kim* was, in my opinion, the best thing he wrote.

There was no Suddar Bazaar at Chakrata, but there was one on the hill at Kilana, which had open shops with the shopkeepers sitting cross-legged on a slightly raised flooring. At night the shops were closed by wooden shutters which fitted in the flooring and in the roof. It generally took a man half an hour to buy a tablecloth or any other article of over ten rupees in value. Much bargaining would have to be done, as the shopkeepers always asked for three or four times its price when negotiations started. At the end of the month we visited the Bazaar. Before entering it from the Chakrata side we had to climb some very wide, steep, stone steps and then a steep hill: on the other side of it were the bungalows of the British troops drawn from different units on the Plains. We approached an old shopkeeper who, with his long white beard, looked like a desert sheikh. We told him that we wanted to purchase a couple of good tablecloths. He produced quite a number of them, to one of which the Prayer-wallah and I took a fancy. We inquired the price of it.

The old shopkeeper eyed the both of us solemnly for a few moments before replying and then said: "Sahibs, this plenty good tablecloth and plenty dear. Last week Officer Sahib gave one hundred rupees for one tablecloth the same. But to private Sahib me sell for sixty rupees."

The Prayer-wallah promptly offered him fifteen rupees. I said that I would give the same price for a similar one.

The old shopkeeper eyed the both of us as before, saying: "You make plenty fun, Sahibs. You go another shop, have plenty fun there."

We turned on our heels and strolled away; we knew very well that he would not let us go without bargaining, because we had let him see that we had plenty of money.

We were not three paces from the shop when he shouted: "Come back, Sahibs, me make fair bargain with you. You no rookies, you very old soldiers, Sahibs, and plenty hard men." He now produced an exactly similar tablecloth, which he offered us at fifty rupees, saying that there wasn't a shop in the Bazaar that had even one of the same quality. After twenty minutes bargaining the price had come down to sixteen rupees, at which it remained for another ten minutes, the shopkeeper swearing by the Beard of the Prophet that he would not take a pice less.

The Prayer-wallah now tried what a little swearing would do. He opened fire with his battery of crab-bat. By the time he had finished, the shopkeeper was trembling like an aspen leaf and gazing at him with an awed look on his face. He double-salaamed three times before saying: "Sahib, Sahib, me mallam (know) you twenty years ago in Connaught Rangers. You young soldier then, now you very old soldier in other regiment, but me mallam you just the same, Sahib." The Prayer-wallah was only four years older than me, but his fierce upturned moustache made him look a bit of a veteran and the old shopkeeper could not believe that any man could have learned to sling the crab-bat to such order in less than twenty years in the East. We got the tablecloths, the first one he had shown us and the other, for fifteen rupees apiece; also Kashmir shawls and silk handkerchiefs, dirt cheap as well. No doubt the old shopkeeper had made a handsome profit out of the sale, but he groaned and swore he had lost many, many rupees over it, and that only to an old Connaught Ranger and his friend would he have shown such weakness.

The Connaughts had a reputation second to none for the way they handled the natives. When they arrived at a new station they soon expounded their views of the race-question and, by the time they left, there were not many natives around that place who even thought privately that they were the equals of white men. The Connaughts were strong believers in the saying that what had been conquered by the sword must be kept by the sword; but not being issued with swords they used their boots and fists to such purpose that they were more respected and feared by the natives than any other British unit in India. It must have been a glad day in the bazaars in Northern India, shortly after the war, when the Connaughts mutinied as a protest against the same sort of methods being used by the Black and Tans among their own families in Ireland, as what they were using among the Indians; and the Regiment had to be disbanded. Every British unit I came across, though perhaps not so brutal in their methods, handled the natives in the same manner. Drafts of recruits were inoculated by the older soldiers in the way natives should be treated, and they in their turn inoculated other drafts. Some of the natives wore boots, which they were very proud of, but we never allowed one of them to enter tent or bungalow except with bare feet the same as they had to remove their shoes before entering a mosque. They were allowed as far as the verandah with their boots on, but if they had to enter the room they must pull them off and leave them on the verandah until they came out again. Natives who entered tents or bungalows improperly dressed in boots were immediately booted out, and they got such a booting that they never repeated the offence. They had to realize that they were our inferiors; and while they did so all was well with them, and with us, and with the whole land of India, where they outnumbered us by a couple of thousand to one.

For those who wished to learn Hindoostani properly, apart from the crab-bat, there was a special native schoolmaster, working under the orders of the Regimental Schoolmaster, who gave classes in Hindoostani and also in Persian. The men who attended this class were mostly men who intended to take on a civil job in the East after leaving the Army. There was a government examination in these languages and a soldier who passed either of them was awarded a bounty – I think fifty rupees for Hindoostani and one hundred and fifty for Persian. Not many men attended these classes, but I was friendly with two who had won certificates. They were both unusual men and I shall tell their stories later. There were often decent jobs given in India for ex-soldiers of good character, mostly jobs on railways or as foremen in construction works and other places where a sharp eye had to be kept on a large number of natives engaged in some simple task.

About the middle of July the monsoon arrived. For five or six weeks there was nothing but heavy rain and mist so thick that it was seldom possible to see more than ten yards ahead. When occasionally the mist did clear and it did not happen to be raining, distant objects seemed a lot nearer than what they did before. The Plains, which were seventy to eighty miles away, looked like a vast sea with islands dotted here and there in it. On moonlight nights, if there was a break in the mist, it was a lovely sight to see the rolling white cloudbanks in the khuds below us, and it gave us a strange feeling to be living above the clouds like the Gods of old. The mist made our clothes damp and clammy, so a daily ration of charcoal was issued to each room, with fire-buckets and light wooden structures for drying clothes on. We generally put these wooden structures in front of the log-fire in the room and used the charcoal in fire-buckets for cooking up porridge, made with Quaker Oats, which was warming for the blood. The helio station working in conjunction with the one at Landaur was packed up for the time being and little signaling work could be done in the open. We were all relieved when the monsoon had passed over: we had glorious weather again, although a little colder than before. During the last two weeks in October the Snowy Range looked as if it were moving nearer to us, but it was only the snow falling at the foot of the peaks and daily advancing nearer to us that caused this illusion.

NORTH CHINA VETERANS

There were three old signallers in my room who had taken part in the suppression of the Boxer rising in North China. Two of them, called Robb and Big Jim, were always swopping yarns about it but the other very rarely spoke about China. I think he had something on his mind in connection with it. From what I gathered of their talk the trouble had started with a big popular movement against foreigners which was encouraged by the Dowager-Empress of China. The Boxers were a secret society. Three English missionaries, the German Minister at Pekin and a Japanese Legation official were murdered, together with a vast number of Chinese Christians. In Pekin the Foreign Legations were besieged early in June 1900 and held out against hordes of Chinese until their relief two months later, losing heavily during the defence. All the leading powers in Europe, and also Japan and the United States, took action, sending contingents to the assistance of their residents in China. The first relief force of three thousand men of various nationalities was beaten back in June on the way up from Tientsin, where the foreign settlements were in the meanwhile attacked. Our Battalion was stationed at Hong-Kong and four companies were immediately rushed up north on board H.M.S. *Terrible*. After passing Shanghai the cruiser ran into a deep fog which delayed them for hours, so that when they arrived at the Taku Forts they found that these had just been captured by the Allied fleet.

As usual there was a lot of international jealousy and it was some time before an expedition could be got under way. The British had to bribe the Japs with a grant from the Treasury before they could send adequate troops. At last there were enough men got together to storm the native city of Tientsin, which was done in the middle of July; and early in August the Commander of the British contingent, General Sir Alfred Gaselee, managed to get the Russians, the Japanese, the Americans and small contingents of other nations on the move towards Pekin. Our four companies were the only regular British infantry represented: the rest were Sikh and marines.

At Tientsin there was a tidy bit of looting and when our fellows started out for Pekin many of them had their pockets full of Mexican dollars and the unwieldy silver ingots, called sycee, which were the chief currency in China at this time; many in addition had dropped a lot of dollars down their trouser-legs where they rested on the top of their puttees. Before the first day's march had been completed they had thrown most of their wealth away and were cursing the Chines authorities for not adopting a sensible currency, gold sovereigns and half-sovereigns, and five pound notes. After a few days' advance they came upon the bodies of men killed in the previous attempt to relieve Pekin. Some the British marines were horribly mutilated: their heads been cut off and placed on their chests and their privates cut out and fixed in their mouths. Photographs were taken of the bodies, which were later sold at Hong-Kong. Robb had two of these photographs, and the corpses looked a sight. He offered me one as a souvenir, but I would not have it. He could easily have sold the two for a lot of money but he was never short, because of working a Crown and Anchor board: he still had the photographs when he left the Battalion.

They had a few minor engagements on the way up to Pekin. In the course of one of them a signaller of ours named Jackson, who was afterwards our signalling-sergeant at Agra, won the Distinguished Conduct Medal for gallantly getting up on a mound and waving a flag under fire, to attract the attention of a Russian field-battery that had mistaken our men and the Yanks for Boxers and was shelling them. Pekin was reached by forced marches on the evening of the 13th August.

The Russians tried to get ahead of the other allies by attacking the same night, but they bungled the job and lost heavily without getting into the city. The Japanese were also repulsed the next morning, and the Americans were held up by the Russians getting across their path. But the British got in by way of an unguarded water-gate and were soon at the British Legation. The wealthiest business part of Pekin had already been destroyed, in the course of the siege of the Legations, by the Boxers setting fire to a foreign drug-store, the flames of which consumed most of the silk-shops and jewellers' shops and so on. The Boxers, in attempting to capture the British Legation, had also set fire to the Chinese University of Pekin where all the most valuable books and archives were stored. But there were plenty of pickings left, especially a day or two later, when the allied troops entered the Forbidden City, from which the Dowager-Empress had only just escaped with the Emperor, and began looting the famous Summer Palace.

The orders on the first night were that the city would be given up to loot for twenty-four hours. Our officers said that all loot must be taken to a certain building for collection and redistribution. But nothing was taken there. The difficulty, according to Robb and Big Jim, was not so much the looting as the knowing what was valuable, and finding valuable things small enough to carry off in a haversack. Some were lucky enough to get hold of genuine black pearls, which they sold when they got back to Hong-Kong for decent prices. Many filled their haversacks with silver dollars: they would be unable to carry them when they were on the march but shared them among their comrades who had not been too successful in their looting. The luckiest men were those who had a friend on the Transport, and several of the officers who had a keen eye for curios managed to get away some priceless ornaments in the Officers' Mess-cart.

During the evening when the looting was at its height Robb and another man entered the house of an old Chinaman whom they found shivering with fright in a corner; not coming across anything of value they concluded that he had his valuables concealed somewhere. They made him understand that if he unearthed his treasure they would not kill him, but if he was obstinate they would. The old Chinaman shook his head, saying that he was a very poor man and that he had no valuables to hide. Robb's mate lost his temper and was just about to drive the bayonet home when Robb stopped him saying: "No, not that way! I'm going to shoot him. I've always had a longing to see what sort of a wound a dum-dum will make and by Christ, I am going to try one on this blasted Chink!" He raised his rifle and shot the old man through the head. He fell dead at their feet. To satisfy his longing Robb examined the back of the head where the bullet had come out and what he saw caused him to remark: "Christ, the dum-dum has blown out the back of his bloody nut!" Dum-dums were soft-nosed expanding bullets, called after the village of that name near Calcutta, where there was an arsenal. At an International Red Cross conference in the previous year it had been decided that dum-dum bullets should be reserved exclusively for warfare against savages – to give them an idea of the advantages of civilization, I suppose. But ordinary bullets could be converted into dum-dums by slitting the nickel casing at their noses, and this is what Robb had done with one of his. In the trenches during the War snipers sometimes used dum-dums – to avenge the death of a pal, mostly. But except at point-blank range a home-made dum-dum would not fly true to a man's aim. There is this to say in favour of the dum-dum: that it will stop a charging man dead in his tracks as an ordinary rifle bullet cannot always be trusted to do. And compared with mustard and other gas a dum-dum is a very humane means of killing one's enemy.

Robb often used to say that if he had known the value of some of the jade vases he had laid his hands on, and if he had been able to carry them on the march he would have been a rich man for the rest of his life. Very nearly all the men who had served in North China retained souvenirs of that show. Big Jim had a very heavy gold watch of peculiar design and make; he had been offered £25 for it but he refused to sell. The original owner must have been a wicked old rascal and a star-turn as well. The watch was a perfect timekeeper. On one side of its face were the hands and numerals, on the other side were two beautifully engraved figures of a man and a woman who were connecting at every tick of the watch.

In September, when everything was calm again at Pekin, there arrived a force of over twenty thousand Germans. They had been addressed by the Kaiser beforehand in the famous speech in which he told them to emulate the Huns of old and have no mercy on the Chinese barbarians who had murdered their Minister. They seemed disappointed that the fighting was over, and the other contingents who had done all the fighting resented the way they began throwing their weight about. Their commanding officer was Field-Marshal Count von Waldersee and he assumed the supreme command of the Allied Forces. No more looting was now tolerated, shops were opened again, and our men were allowed in the city with special passes, the walking-out order being a belt and swagger-cane. Six of them one morning happened to drop in at a Chinese place which sold beer. There was a bagatelle-table in the bar and after a few drinks a couple of them began to play a game at the table. This had only been in progress a few minutes when in walked seven or eight German soldiers carrying sword-bayonets in their scabbards. The leader of them said "Ach" and spat over the table. One of the players retaliated by poking him in the face with the cue he was using, injuring his eye. He started to yell and curse and while his comrades gathered around him one of our chaps knocked in a small window at the side of the room and six of them jumped out of it before the Germans noticed what they were up to.

Our men knew they were in for a scrap, but they had no intention of running away. As they had entered the place they had noticed a small plantation of strong young fruit trees outside this window, only recently planted. These they easily pulled up for use as weapons against the sword-bayonets of the Germans.

The Germans did not follow them through the window; they rushed out of the place with their sword-bayonets drawn. They were out for blood for the injury to their comrade's eye. Our chaps were out for the same, for the spitting insult. The Germans were at a disadvantage with their sword-bayonets: these were not long enough to cope with the young trees, which were about five feet in length. They had a difficult job to get to close-quarters to drive them home, so had to be content with aiming cuts at their opponents' heads. Our chaps managed to disarm three of them and batter them to the ground with the trees, using the root-ends as clubs. By that time four other Germans who happened to be passing had joined in the fray. At last two armed pickets appeared on the scene, including one of our own regiment, and put an end to the fight, but not before four of the Germans were dead and two seriously injured. Four of our chaps had also to be carried to hospital, bleeding from bayonet-cuts. The following day the surviving Germans were brought along to the hospital to identify their wounded assailants; which they did. But nobody seemed to know what to do with these men on their discharge from hospital: their crime was an international affair. They were not court-martialled but were kept as prisoners at large for about two months. Then orders came through that the officer in charge of the British Legation Guards should try them. This happened to be Major Sir Horace Mahon, Bart., of their own battalion. The gallant Major gave them forty-eight hours' cells, and told them unofficially that he had given them this punishment for not having made a job of it and killed every damned German in the gang. He said that dead men tell no tales. I give this story for what it is worth: it is how I have always heard it told.

During the time they were in Pekin our chaps sometimes had to escort between fifty and a hundred prisoners to the place of execution before breakfast. The executioner used a double-handed sword, with which he cut their heads off, while they knelt, like greased lightning. Big Jim had been present at some of these executions, which so upset him that he could never eat a decent breakfast after one of them. He must have a different stomach from what the old Meerut Bacon-wallah had: he would have given the half of his piggery to be present at such a spectacle and no doubt would have breakfasted off a sirloin of beef and ten pound of fried onions to celebrate it. It was the way the blood spouted up.

I must tell what happened to Robb. Men who were serving twelve years with the Colours and had completed nine years of their engagement, six of them abroad, could claim to be sent back to the home establishment. Robb had engaged to do his twelve years with the Colours and intended, when he completed eleven years, to re-engage for another ten. Twenty-one years would qualify him for a pension of a shilling a day for life. We were at Agra when Robb applied to be transferred to the home establishment. He was sent back and joined the First Battalion in Ireland. There on the Curragh one day he went with a lot of other men to bathe in a large pool. He dived into the water, where no doubt the ghost of the murdered Chinaman was lurking in wait for him. He did not come up from his dive and when his chums came to miss him they found him lying at the bottom near a big stone on which he had apparently struck his head. They pulled him out half-drowned and unconscious, with a big bruise on his head, and carried him to hospital. The injury left half of his body permanently paralysed. I never heard what became of him in the end but the doctors did not hold out much hope for his recovery.

The Colour-Sergeant of my company, who was called Bottle, fancied himself as a big-game hunter. He was determined that he would bag a cheetah before the Battalion marched back to Meerut. One fine moonlight night he tied a young goat to one of the goal-posts on the Bottom Square. In the course of that afternoon he had had the ground around it whitewashed, so that he would have a better view of anything that approached the goat. Just behind the goal-posts on a rise in the ground was a small fir-copse. He chose a spot on the fringe of it where he waited patiently but eagerly for the arrival of a cheetah. Two hours passed. The goat bleated pitifully but no cheetah put in an appearance. However there was one somewhere about. Bottle heard it cry several times, and it seemed to be coming nearer. Then for a long time there was absolute silence. He was afraid that the cheetah had become alarmed and moved off. A mass of dark clouds obscured the moon for a couple of minutes and, when it passed, Bottle rubbed his eyes in amazement. Not only was there no cheetah, but the goat had disappeared as well, and he had not heard the least cry or scuffle. The cheetah must have been an old campaigner, who had lain crouching behind some boulder on the side of the Square until the moon should be obscured. As soon as that occurred he must have made a few lithe springs towards his prey and bitten it in the throat, severing the cord as he did so. When Bottle related his story the following morning at breakfast in the Sergeants' Mess they all chaffed him unmercifully. They said that he had been lucky that the cheetah hadn't pinched his rifle and ammunition as well as getting away with the goat.

I don't think that any battalion in the British Army had a better lot of officers than what we had, on the whole, during my service with them. They took a keen interest in the men and fostered all kinds of sport. Most of them belonged to the landed gentry of North Wales and the Welsh Border. They were strict disciplinarians, in the sense that if they gave orders they expected unquestioning service, but they were far from treating us with contempt. The case was one of mutual trust in military matters and matters of sport, but no social contact. If a man wished to speak to an officer on some private matter he could only do so by first approaching his Colour-Sergeant, who would take him in front of the officer and remain present during the conversation, in case the man should use a disrespectful tone or make some irregular request which the officer, if he was young and inexperienced, might be tempted to grant. The gulf between the Regimental Sergeant-Major

115

and the youngest officer in the Battalion was equally great, if not greater than the gulf between the Regimental Sergeant-Major and the youngest recruit. Even when our R.S.M. was given a commission as lieutenant, and appointed Quartermaster to the Battalion, he did not on that account regularly dine in the Officers' Mess. He was merely invited once a week on Guest Night. We had no other ranker officers at this time in the Battalion, though the aged Colonel of the Regiment, Sir Luke O'Connor, v.c., had won his commission during the Crimean War, while he was a sergeant. (Sir Luke's commanding officer in the Crimea was a certain Colonel Holmes, who had served with the Regiment at Waterloo, and whose grandson "Chopper" Holmes, who joined us later as a recruit officer, rose to command the First Battalion during the War. There were also cases of men in the ranks who claimed to have had fathers and grandfathers serving in the Regiment, which went in strongly for tradition.)

The only link that joined officers and other ranks in a closer human manner was Free Masonry. I could not write about out this in detail if I were a Mason, because of the oath of secrecy; and I cannot, because of ignorance, since I am not a Mason. But I happen to know that, for instance, "Papa" Yates, who managed the Regimental soda-water factory at Agra and later in Burmah (a profitable business, much discussed by the troops, who were always smelling rats), was one of the Craft, and attained a high degree in it. He was a Colour-Sergeant, but acting as Orderly Room Sergeant at this time. Not long after I became time-expired he was made Quartermaster-Sergeant, a rank which he only held a short time before being gazetted Lieutenant-Quartermaster, serving as such all through the War. This promotion caused a lot of comment, because it was not due to him by seniority. But he made a good Quartermaster. As a Mason he must have had the privilege of calling his Commanding Officer "brother" while a lodge was in progress, which seems a strange state of affairs. In theory, I am told, any man of good moral character who believes in a Supreme Deity, and is loyal to the King, and can collect a couple of sponsors and the necessary initiation fee, can become a Mason; and I have been told, too, that no officer who is unacquainted with the grip and other high signs can get anywhere in the Service. But in practice I know that in my time it was impossible for anyone under the rank of full sergeant to become a Mason. I know, too, that any man who attained full-sergeant's rank and intended to make the Army his profession did not hesitate, but immediately began looking around for sponsors. Shortly after the Great War, by the way, I joined the "Antediluvian Order of Buffaloes," which is known as the Poor Man's Free Masonry, and is very strong in some parts of the country among those who like their drop of "bitter-gatter" or "juniper-brandy," as beer and gin are called in the secrecy of the Buff Lodge, and who don't mind a sing-song. But I only attended one session, the same as I only attended one session at Sunday School as a kid. "Once a Buff, always a Buff," is a saying that is true of me only in the sense that my name is still, I expect, on the Buff register, and I have never revealed the password or clasp to a soul (the truth of the matter being that I have forgotten both).

After a colonel had been in command of a battalion for four years it was now the rule that he retired on half-pay to make room for the next senior officer. Our Commanding Officer at this time was Lieutenant-Colonel Bertie, who had been awarded the C.B. for his services during the Boxer rising. He was too much of a martinet to be greatly beloved by the rank and file, and punished them to the extreme limits of King's Regulations. About a month before we left Chakrata he had completed his four years; and one morning, just a week before he gave his farewell speech to the Battalion on the Bottom Square, he woke up to find that the plants, flowers and hot-houses around his bungalow, which were his chief interest outside of the maintenance of discipline, had been totally destroyed. He suspected that some of the men whom he had punished with his usual severity at some time or other had done the damage, but he could prove nothing. Personally, I believe that a troop of large monkeys that often prowled around the station during the night had done it all; they would sometimes clamber on the roofs of the bungalows and kick up a terrible row.

The Commanding Officer's speech ran something like this: "Officers, non-commissioned officers and men of the Second Royal Welch Fusiliers: during the four years I have commanded this battalion we have been through much together. We were the only British infantry battalion to take part in the North China campaign, from which we emerged with honour. Only a week ago some despicable hooligans in the Battalion destroyed my plants, flowers and hot-houses; but what they cannot destroy is my good name. Up to the very last minute of my command I shall do my duty to the Regiment and I very much regret that I am leaving you tomorrow." Most of the men in the Battalion were wearing broad smiles by the time this short speech was over; and I even heard tittering noises from one or two of the junior officers who had suffered from Colonel Bertie's sense of discipline in much the same way as what we had.

When Colonel Bertie left, so, naturally, did his wife. She was a striking woman, taller than what he was and a real lady. She used frequently to pay visits to the soldiers' wives in the Married Quarters, though of course the same wide gulf that was fixed between officers and men also separated their women-folk. I often did a grin at some Battalion outdoor function, such as Regimental Sports, to watch the ladies according to their different social classes collect in groups apart from one another: one group of officers' wives with the Colonel's wife in command, another of senior N.C.O.s' wives with the Regimental Sergeant-Major's wife in command, and then the wives of the sergeants, corporals and privates, each group parading separately. It was class distinction with a vengeance. Curiously enough, the best-looking woman of them all and the one who wore her clothes in the smartest manner was the wife of an undistinguished old private. But her superiors in class were too well-bred and sure of themselves to show any irritation at her superiority in other respects. There was an old rhyme, to the tune of the Officers' Mess-Call:

Officers' wives have puddings and pies
But sergeant's wives have skilly,
And the private's wife has nothing at all
To pack her poor little belly.

It must have dated from very early times, like the bugle-calls themselves. Skilly was a gruel that they now issued only to convicts, and the private's wife who was on the strength drew good rations of bread, meat and potatoes, the same as a sergeant's – she and her husband could even afford native servants. (It must always have been a bit of a come-down for her when her husband was time-expired or transferred to the home-establishment, and she had to get down on her knees and scrub floors again, or take her turn at the wash-tub.)

A few married women were left each year on the Plains at Agra during the summer, but those that had young children were generally ordered to spend the whole of the summer at the Hills. Children born in Barracks were referred to as "barrack-rats": it was always a wonder to me how the poor kids survived the heat, and they were washed-out little things. It was rarely that a woman was sent to a Hill-station without her husband, but if that happened and if she was young and pretty she would be sure of an invitation to all the dances and social functions held by the N.C.O.s there. She would have to be very level-headed and virtuous to resist all the temptations that came her way, especially if her husband did not join her with the half-time relieving party, about which I shall have something to say later. Six months in India is a hell of a long time for husband and wife to be parted.

Most of our married women were very respectable. (I believe that when an application was made to the Colonel for a man to be married on the strength the Colour-Sergeant of the company was asked to report privately on the woman's respectability before this was granted.) I can only remember two of them that had regular fancy-men. Everybody in the Battalion knew about this, but it was a long time before one of the husbands found out that he had a rival. When he did, his wife was confined to her house for some days with two lovely black eyes; but he spent a few days longer than this in hospital, with stitches in his head where she had broken a jug on it. Whether the other husband winked his eye at his wife's carrying on, or whether he was too thick to find her out, was a subject of long dispute with us. In any case, the two of them seemed to get along well enough together. If there was ever a serious squabble between a married couple the Colonel's wife would step in and do her best to make peace between them. Mrs. Bertie was an excellent peace-maker: nobody could continue cursing and throwing things at each other in her presence. I never heard of a divorce among the rank and file and cannot imagine such a thing happening; a murder would have been far more likely. I knew two men who married prostitutes, but this was after they left the Battalion. The deed was looked upon with loathing and contempt by their former comrades who, although they went with prostitutes themselves, would sooner have died than marry one of them.

ARCHIE, AND THE
ASSISTANT COOK

On the twenty-ninth of October we began our march back to Meerut under the temporary command of Major Beresford Ash. This time I was not with my company, but at the head of the Battalion; for I had recently been taken on the strength of the Regimental Signallers. The head of the Battalion was always our position on the line of march. On ceremonial parades the Goat and Goat-major had this place of honour, immediately followed by the Pioneers, who had the privilege that no other Pioneers had in the British Army, of wearing white buckskin gauntlets and aprons, which looked very smart; then came the Band and then ourselves, with slung rifles and small furled flags, and then the leading company with their rifles at the slope. Ceremonial drill, especially when the Regimental Colours are carried, has about as much connection with ordinary soldiering as religion has, but it is grumblingly accepted by the men as a necessary evil and is said to assist recruiting.

The other troops at Kilana were also under orders to move, so the shopkeepers and the girls in the brothel were packing up for the Plains. The Suddar Bazaar would not come to life again until the first week in the following April. A winter section of twelve men under the command of a sergeant was left at Chakrata. One of these men told me later that they had been practically snowed up for three months of the winter and that most of their time had been spent in clearing snow from around their bungalow. If no snow had happened to fall in the night they usually found in the morning the tracks of a bear that had come right up to the door of the bungalow. But they never managed to get a shot at it.

At Meerut we were met by our new Commanding Officer, Lieutenant-Colonel T. Lyle, D.S.O., who had won a name for himself as a leader of Mounted Infantry in South Africa. Tommy Lyle, as he was called by the men, was very popular during his four years of command. He did not believe in too much ceremonial drill: the ability to march and shoot straight were the two chief things that he required from the men under him. There were plenty of field-days at Meerut, where the size of the garrison allowed schemes of attack and defence to be carried out over a large extent of the country. When there was no field-day, ordinary company-drill was carried out up to noon, with occasional Battalion parades taken by the Colonel or the Adjutant. Sometimes the Regimental Sergeant-Major would give the Battalion an hour's drill in the afternoon, but generally speaking one was free in the afternoon and had the choice between a nap, a ramble, and a gamble.

Some men did not take to Army life as easily as I myself did, and most of the other men, and a few were so fed up that they would do anything to get out of it. At home I have known men steal something costly out of a shop, or smash a window, in the hope of a civil conviction which would lead to their getting discharged. A discharge with ignominy to these men was better than no discharge at all. After they had been punished by the civil authorities they could expect a strong sentence by the military authorities, in addition, up to six months' imprisonment. The risk they ran, however, was that they might get both sentences and do their full time in clink, and then be retained in the Army, into the bargain, because hitherto they had borne fairly good characters. Others tried to get invalided out of the Army on grounds of ill health, but it was difficult to make oneself ill enough to do this without inflicting a permanent injury. One man pretended that he had gone stone deaf. He started off with not hearing words of command on the parade-ground, and then, when he was made a prisoner, not hearing his company officer's sentence. He was sent to hospital for observation and kept up his pretence for ten days. Finally the medical officer pretended to believe him and then got an orderly to drop a rupee on the floor behind him when he was off his guard. Thinking it had fallen from his own pocket, he turned round and picked it up. He was immediately made a prisoner again and given fourteen days' imprisonment for the serious crime of "going sick without a cause."

Two more men, whom I knew very well, pretended that they had lost their mental balance. They hoped to be sent to a military insane asylum. Once there, they reckoned on being discharged in a few months' time and certified sane enough to return to civil life. One of the two started boldly enough but his methods were too ordinary to have even deceived his old grandmother. He stuck straws in his hair and went about grinning and stalking flies, which he would try to catch with his bare hands, and then he gave out that he was Field-Marshal Earl Roberts of Kandahar. After a few terms of imprisonment he threw up the sponge. But the other, who was called Archie, was a sticker. He had joined the Army in a fit of despair over the young lady with whom he had been walking out, who had chucked him and taken up with a soldier. It was Archie's idea to be sent to the South African War and win a posthumous V.C., so that she would be sorry for the manner in which she had jilted him. But after he had been in the Army for a time he forgot his broken heart, and the South African War ended, and he wanted to return to civil life.

It was while we were in Jersey that Archie began to "work his ticket," as it was called. An Adjutant's Parade was the first occasion. His company, which was the leading company on the Square, had already fallen in, the roll had been called and the orderly-sergeant had reported Archie absent, when he came strolling out of his room, trailing his rifle behind him, with a far-away look in his eyes. He fell in on the left of his company, just as the company officer began to inspect it. The Adjutant, who spotted him, rubbed his eyes in amazement and certainly the way Archie was dressed would have made a cat laugh. On his red jacket he had stitched a dozen lids of Day and Martin's Soldier's Friend, together with metal-polish tins, all of them highly polished. Tied to the back of his braces and hanging over his backside was a frying-pan. The Adjutant was too astonished to say a word until Archie was about to be marched to the Guard-room under escort. Then he roared: "Bring that damned lunatic in front of me." When questioned as to why he had appeared on parade improperly dressed, Archie assured the Adjutant that he was properly dressed. He said that he was entitled to the decorations and medals he wore on his breast, having won them during the years he had served with the Emperor of Abyssinia's army. He said that the large decoration he wore on his backside was the most coveted honour in Abyssinia; when the Emperor decorated him with it he had also promoted him to full general. He said that the generals and the princes became jealous, and if he had not left the country when he did he would have been dead in a very short space of time. He said that when he arrived back in England, the only proof that he had that he had been full general in the Abyssinian army was his decorations, which he was now wearing and which he was very proud of. The Adjutant ordered the escort to take him to hospital, but after he had been there a week the medical officer came to the conclusion that he was perfectly sane, and Archie was sent back to the Fort under escort. For making a laughing-stock of the King's uniform and pretending he was barmy he was lucky enough to get the light sentence of fourteen days' cells.

He still acted strangely after he came out of the cells. He was determined to leave the Army by some means or other and, like other men who tried to work their tickets, he did not have the necessary twenty-one pounds to buy himself out. He would have deserted but he knew that if he went back to his relatives he would soon be arrested as a deserter, which would mean six months' imprisonment, probably without a discharge at the end to console him for his hardships. He decided to stick to his original plan of action. In India his manner became stranger than ever. He used to have long interesting conversations with himself, mostly about love or Abyssinia, and was twice sent to hospital for observation. He was not punished any more, because the doctors were undecided as to whether he had lost his mental balance or not. A lot of us believed that he was really up the loop from having played at it so long. On our march back to Meerut we stayed one day at a place where there was a magnificent temple on the banks of a large deep lake. That afternoon quite a number of us were enjoying a swim in the lake and Archie, who could not swim a stroke, sat on the edge of the lake watching us. Some time later, when most of us had left the water and were just beginning to dress, I heard a man exclaimed: "By God, that man must be an expert, otherwise he wouldn't dive in from as high as that!" I looked around and was surprised to see Archie stripped and standing on the top of a high pillar of stone on the edge of the lake. I shouted to him not to be a fool, but at that moment he made a wonderful dive, going in so straight and making so little of a splash that the men who did not know him very well uttered a cry of amazement and said: "That fellow Archie must have been a professional high-diver in civil life!" But the rest of us, who knew him better, dived in and fished him out half-drowned. This was a further escapade to be reported to the doctor who was attached to the Battalion on the march.

Archie's final stunt was a masterpiece. The Divisional Sports were being held in a few weeks' time and he entered for every running event from the hundred yards to the mile. He refused to be assisted in his training, which he did, so he said, about an half an hour before twilight every evening. Late on Sunday evening, on the day before the Sports were to be held, Archie left the tent, saying that he was going out for a final spin on his secret training-ground. We who knew him were wondering what new scheme he had evolved for working his ticket. We had not followed him before when he was going out for his spins, but on this evening we thought we would. We kept about one hundred yards behind him. He stuck to the main road after leaving the

Camp until he was about twenty yards from the entrance to the Protestant Church. There we were surprised to see him cut across country and disappear among some trees and high tropical plants at the back of the church.

The shadows of twilight were falling as we arrived at the entrance to the church, where all the best society of Meerut attended evensong. After a little discussion we decided to enter the grounds, and get among the trees at the back to see what Archie was up to. But just as we opened the gate the congregation began to file out and collect in little groups here and there, as the custom is all over the civilized world, gossiping together about fashions or the sermon. Suddenly a man with a pair of running pumps on his feet but otherwise as naked as the day he was born jumped out from behind the plants and begun running round and round the church with the speed of a hare. It was Archie. Some of the ladies screamed, others did their best to close their eyes. I expect that the full-blooded ones who had old and decrepit husbands closed only one eye and gazed with the other in rapturous admiration at this nude athlete. Archie was physically handsome in feature and limb and old Mother Nature had been kind to him in many ways. For a few moments the ladies' esquires were too astonished to do anything, and it was the same with us. He completed two laps and was halfway around the church on a third one before we burst into the grounds, shouting that the man was a lunatic. We caught him and rushed him behind the plants out of sight. The three of us now thought that he was really up the loop. He did not seem to realize that he had done anything out of the ordinary and said: "Well, boys, do you think any man has a ghost of a chance against me tomorrow? You'll see, I'll cake-walk every event I have entered for."

An officer of another regiment, who had been in church, ordered us to conduct him to hospital and in less than a fortnight he was on his way down to Kalabar under escort to be interned in an asylum there. Later he was transferred to a military asylum at home. His last words to the escort were: "Well, so long, boys. I'll be thinking of you when I'm back in Blighty. I am supposed to be barmy, and so I was to join the Army. But, one thing, I'm not half so barmy, and never have been, as those barmy bastards who still have to do six or seven years in this God-damned country. You'll be doing me a favour if you convey to them my deepest, heartfelt sympathies." Within twelve months we had news that he had been discharged from the asylum and from the Army, and that he had an excellent job in his home town and was happily married to a young lady who, he said, was worth a hundred of the one for whose sake he had behaved in such a rash manner.

The following spring when marching back to Chakrata we pitched camp at a place called Muzaffarnagar. That afternoon a number of us were watching a native groom exercising a beautiful white Arab horse belonging to one of our majors. The major was there too, with a few other officers. He had only just bought the horse. A large crowd of natives from the place were also watching in the background and all admiring the horse, which was really something worth admiring. Suddenly out of the crowd sprang one of the natives. With one swipe of his arm he knocked the groom flying. At the same moment he seized hold of the leading-rein, took a flying vault and in another second, if he hadn't been unlucky, he would have been on the horse's back and off like the wind. His bad luck was that one of our Regimental police, named Portman, happened to be present. As the horse-thief sprang towards the groom, so Portman, a man who was never taken by surprise, sprang too. He caught the thief in the middle of his flying vault with a smashing blow on the head from a heavily loaded stick. The thief fell senseless to the ground. The crowd of natives melted away in a few seconds but it was quite an half an hour before the horse-thief recovered his senses even with the help of a bucket of water. He was then spread-eagled outside the Guard-tent and by the time the native police arrived to take him away he was more dead than alive. He turned out to be one of the most famous horse-thieves in the whole of Northern India; for years the police had been trying to catch him, but he always worked on his own and so nobody could be found to give him away. He was never known to steal a horse during the night and in the course of his career he had stolen dozens of first-class horses in exactly the same way as he had attempted to steal the white Arab; nor could the police ever find out where he sold them. They said he was far cleverer than any rifle-thief. Muzaffarnagar was well-wooded and, once out of sight, the thief, who knew the country well, would never have been caught.

Our next station was Agra, about a three hundred miles' march from Chakrata. On the day we arrived at Meerut, the Battalion had to do ten days' manoeuvres before continuing with the march, the Goat died. I think it was old age, he had been with the Battalion since he joined it eight years before, at Malta, or it may have been Crete. Every man in the Battalion was genuinely sorry that the wicked old rascal had gone West, but the vegetable-wallahs were mighty relieved: they did not say much but they walked around the camp with a jauntier air than what they had done when he was alive. He was buried underneath the big tree where they hanged the rebels during the Mutiny. If he could have spoken I have no doubt he would have chosen the same spot for his last resting-place. It was only a few yards from the little tent he used to stay in, from which he had many times broken loose to trail a vegetable-wallah.

The old Bacon-wallah was present at the burial and informed us that if his family carried out his wishes he also would be buried under that same tree. A large cross was put on the Goat's grave, giving full particulars of his service. Three years later I paid a visit to Meerut and found the grave in beautiful condition; one of the men of the Battalion that was encamped there told me that more work was done on the Goat's grave than on all the soldiers' graves in the cemetery put together. A kind of unwritten law existed among the troops in India that the graves of regimental mascots or of men who had happened to be buried outside cemeteries should always be well cared for; the cemeteries were looked after by the natives who were working there and were often allowed to run down and get overgrown with jungle vegetation. The old Bacon-wallah was very much alive on this visit, and pronounced a eulogy on the Goat in these words: "He was a Royal beast and I held him in sincere admiration. He could not read or write, but neither can many of us, and in one respect he put every man of his regiment to shame: that Goat knew better than any one of you how the natives of this country should be handled." I never again visited Meerut after that, but I have often wondered whether the old Bacon-wallah is buried under the tree – even if he had been granted permission by the authorities it is extremely doubtful whether his children would have allowed him to be buried alongside the Goat, though the Goat was so clearly his kindred spirit.

In the first week of December 1904, we relieved the South Staffords at Agra. Some time later the Viceroy, Lord Curzon, visited the city and stayed at the Government House, where we found a guard over him. He looked in very poor health. He had, I think, come to see that everything was in apple-pie order for the forthcoming visit of the Prince and Princess of Wales – now King George[2] and Queen Mary – who were going to spend a week at Agra. He had already completed five years as Viceroy, but had been reappointed for a further term of office. When he retired in 1905 the huge unanimous sigh of relief that went up from the rank and file throughout India must have caused quite an atmospheric disturbance. Lord Kitchener, who late in 1902 had been appointed Commander-in-Chief of the Forces in India, was very popular with all ranks of the Army, and it was common gossip even in the native bazaars that Curzon Sahib and Kitchener Sahib were very bad friends and that India was not big enough to hold the both of them. Former Commanders-in-Chief of the Forces in India had been obliged to carry out orders that were issued to them by the Viceroy's Council, whether they agreed with them or not. Lord Kitchener kicked against this principle, which was one

---

[2] I have just had news of the King's death and it would be useless to pretend that it has caused me genuine grief, like the death of a near relative.

of the chief differences between him and the Viceroy. The two signallers who accompanied the Colonel on large field-days at Meerut and Agra used to hear a lot of news about the high and mighty ones, especially when he was conversing with senior staff-officers. They reported that there was great jubilation among the staff when Lord Curzon went.

The garrison at Agra consisted of ourselves, one battery of Field Artillery, one battery of Garrison Artillery at the Fort, and one or two native regiments. Our Barracks were three miles from the city, and just outside the city was the Fort, where each company in turn did a month's duty. Between the Fort and the Barracks was the civil cantonment where lived the white men and half-castes who had Government jobs. There was a bell always ready to be sounded as a warning in case of a native rebellion or riot: when they heard it they were to leave immediately with their families and take refuge in the Fort or at the Barracks. I believe that every civil cantonment in Northern India had the same arrangement, which dated from the Mutiny. The Fort had formerly been the residence of the old Mogul Emperors. Akbar, the greatest of them all, had built it and had held his court there. I have spent many an hour wandering around the beautiful buildings inside it. Not a stone's throw from it, on the banks of the River Jumna, is the celebrated tomb called the Taj Mahal, which was built by Akbar's grandson Shah Jehan in memory of his favourite wife. In addition to the finest craftsmen of their age, more than twenty thousand men, the majority of them slaves, were occupied for over seventeen years in building it. With the exception of the side facing the river, which from the foundation to a certain height is built of red sandstone, it is all pure white marble. The interior of the tomb with its marble screens and delicate pierced marble-work makes one amazed at the skill and patience of the workmen of old. Although the Prayer-wallah and I were hardened sinners we were also great admirers of all things that are beautiful: on many a night we left the Canteen half cut and journeyed down to view the Taj by moonlight, when it looked three times more beautiful than what it did during the day. Since the invention of cheap winter-cruises, I understand that thousands of globe-trotters go to Agra every year on purpose to see the Taj Mahal by moonlight, having been told by the steamship companies that the sight is something to dream about. But the Prayer-wallah and I found it out for ourselves.

We had not been at Agra a fortnight when we had a new assistant native cook at the Signallers' bungalow who could speak fair English and understand it. He had worked as a lad at a Mission, where he had picked up Christianity and had afterwards been employed at Soldiers' Homes; and was of a robuster build than most of the natives of the Plains. He began his duties on a Monday morning. On the following Saturday morning a certain Private Crickett gave him three eggs to boil, but when he brought the breakfast out on the verandah he handed Crickett the three eggs on a plate, fried, not boiled, and burned an ugly colour into the bargain.

Crickett exclaimed: "You black soor, I told you to boil those eggs. Instead you've fried them until they resemble three blasted bits of cow-dung. For two pins I'd knock hell out of you, you black-faced git." I have forgotten most of my Hindoostani, so cannot reproduce the rest of his remarks.

The assistant cook faced Crickett and said in Mission-school tones: "Sahib, no more today the white Sahibs beat and curse their own brothers. Me made in God's image the same as you. Me know our Master the Lord Curzon Sahib's orders: all white Sahibs who ill-treat us to be punished. You curse me, but you afraid to beat me, and me not afraid of you, Sahib."

Crickett lost his temper so completely at being called the assistant cook's brother that instead of doubling him up with a blow to the body, as he should have done, he delivered a punch that landed on the point of the cheek-bone. The cook dropped like a log to the floor and lay half-stunned for a few minutes with the blood trickling down his face from the gash under his eye.

131

Crickett chucked the eggs over the verandah rails and continued with his breakfast, having to content himself with a khaki patch.

When the assistant cook came to himself he rose to a sitting posture. The blood was now trickling down his chest, and soon as he saw it he turned white and green. Then he to his feet and ran screaming at the top of his voice to Quartermaster's Stores. He made his complaint to the Quartermaster, who was the boss of all the natives attached to Battalion, and said that Crickett Sahib had struck him seven times in the eye. The Quartermaster, seeing that the cook was badly marked, and bleeding like a stuck pig as well, sent the Sergeant to investigate the case and make Private Crickett a prisoner if he was the man who had struck the blow.

While all this was happening, Signalling-Sergeant Jackson was having breakfast at the Sergeants' Mess; the other non-commissioned officers, with the exception of a lance-corporal, finished their breakfasts in a hurry and discreetly vanished, telling us that they had not been present when the incident occurred. The Lance-corporal gave the Sergeant his version of the affair and said that if Crickett had not knocked the cheeky black swine down he would have done so himself. The Sergeant said to Crickett: "You damned fool, why the hell didn't you strike or kick him in the bloody guts? He has plenty of evidence to show now and the Quartermaster will do his best to rub it in for you. All you signallers are on his bad books since one of you gave his pet servant such a severe kicking a week or two ago." This pet servant, who wore boots, had attempted to enter our bungalow without removing them. For months afterwards he pretended that his backside was so seriously injured that he could only manage to limp along by holding it with one hand.

There was no Orderly-Room on the Sunday, so Crickett was tried for his crime on Monday morning. The assistant cook, who was outside the Orderly-Room, looked a ghastly sight. He had not attempted to bathe the gash under his eye and was covered in blood from head to foot. He was a low-caste native and we found out later in the day that he had been down to the slaughter-house where the cattle were killed for the troops and had splashed himself all over with blood, so that he would have enough of visual evidence to show the Colonel Sahib. This was the only evidence he did have; the number-one native cook was in the cook-house at the time and a couple of native sweepers of the bungalow, who had witnessed the affair, had too much respect for Crickett to give evidence against him. Crickett, who had been put back by his company-officer to be tried by Colonel Lyle, had the Lance-corporal and another man as witnesses on his behalf. The cook spun a lovely yarn: he told the Colonel that as soon as he had appeared on the verandah Crickett started to curse him, with blasphemous words. After he had taken the basket off his head, which held the Sahib's breakfast, and set it humbly on the floor, Crickett Sahib had sprung at him and knocked him senseless with repeated blows. His eye had only just stopped bleeding and he had lost so much blood that he had been too weak to wash his body clean from what had gushed from his eye. The Quartermaster also spoke on his behalf, saying that the Signallers did not number thirty men all told, but that he had received more complaints of their ill-treatment of natives than of all the rest of the Battalion put together.

What Crickett told the Colonel in answer to the charge was true up to a point: he repeated the assistant cook's impudent words. But he also swore that the assistant cook, who appeared to be under the influence of drugs, had seized hold of a heavy gridiron which was in the basket and attempted to brain him with it. He had dodged the blow and knocked the man down in self-defence. If he had not done so, he would have been a dead cock. This was a better yarn than the assistant cook's and Crickett was ably backed up in it by his witnesses, the Lance-corporal going so far as to say that the assistant cook should, by rights, be now in jail on a charge of attempted murder.

The Colonel, who was no disciple of Lord Curzon's, listened gravely to this evidence. He then told Crickett that he done quite right in defending himself, but that he had used a trifle too much force, to judge by the depth of the cut under the cook's eye. He was afraid that he would have to punish him. He sentenced him to the almost nominal punishment of three days' confined to barracks. After leaving the Orderly-Room, Crickett remarked that if he was sure that all he served under would be as sensible as old Tommy Lyle, he would spend the rest of his days in India felling bloody natives who claimed to be his brothers and equals. For the next few days the assistant cook, who had lost his job with us, visited the different bazaars around the place in the same state as he had appeared at Orderly-Room. To every native who would listen to him he would point out his damaged eye and blood-stained body, and complain how badly he had been treated by the white Sahibs. At last one of our Regimental police, on duty in the Suddar Bazaar, appeared on the scene while he was pitching his tale of woe to a group of natives. The policeman dispersed the group and, seizing hold of the ex-assistant cook by the clothing at the back of his neck, gave him a severe thrashing with his stick. He then warned him that, if he appeared in any of the bazaars again, the police would half-kill him before handing him over to the native police on the charge of spreading disaffection in the bazaar, which would land him in jail with a heavy sentence. He was never seen about after that.

THE PRINCE AND PRINCESS OF WALES

The second week in December 1905, the Prince and Princess of Wales arrived at Agra. They inspected the Battalion just outside the grounds of the Taj Mahal. The Prince went along the ranks and the Princess gave a few friendly pats to the young Goat which had been sent out from the King's herd at Windsor to take the place of the deceased sinner. Elaborate precautions were taken for their Royal Highnesses' stay at Government House. The Battalion found a guard of one hundred and thirty men, commanded by a Captain with two junior officers and two senior non-commissioned officers under him. We had sentries posted inside and outside the house. A native regiment found a supplementary guard, with rings of sentries encircling the grounds. Between twilight and six o'clock in the morning nobody could pass the sentries unless they gave the pass-word, which was changed every twenty-four hours. The first guard that mounted were at the house four days before being relieved by a force of equal size and composition. It was a very easy guard: instead of sentries doing the regulation two hours on and four off they did two hours on and twenty-two off. For guards of honour and guards like the present one, the tallest men in the Battalion were chosen. It was only on the line of march that Signallers were called upon to mount guard, but on this occasion four of us were chosen because of our height, including the Prayer-wallah and myself.

At three o'clock in the morning I relieved a sentry who was a pukka Don Juan. He was stationed in the corridor where the Princess had her bedroom. Sentries on this post were given goloshes to wear so that they would make no noise when walking along their beat. Don Juan did not want to be relieved. In a whisper he asked the Colour-Sergeant who came round to supervise the changing of sentries whether he could do another two hours' sentry-go. Of course the Colour-Sergeant refused, and I took his place.

Outside the house the Colour-Sergeant asked Don Juan why he wanted to do another two-hours' sentry-go; in the whole of his sixteen years' service he had never before come across a man who wanted to do extra sentry-go. "Well," replied Don Juan, "one of the young white ladies who attends on the Princess is sleeping in the room at the end of the passage. From the sounds I overheard shortly after I came on sentry she had been sitting up late writing a letter home. There was a crackle of paper at last, as if she was folding up the letter and putting it in an envelope, and then she began to undress. I heard her sigh with relief as she unclasped her corsets and got into a dressing gown, and then she brushed her teeth and did a little bit of gargling, and then she started to brush her hair, which she did for a long time with a delightful swishing sound, and then she fiddled about with her toilet waters and her pomatums, and finally she flopped down at the bedside. Then I heard a sound that reminded me of the last time I slept with a white woman, a couple of nights before I left England. It brought back such delightful memories that I would gladly have done another four hours' sentry-go on the off-chance of hearing it again."

The Colour-Sergeant, who was a respectable married man, cautioned Don Juan, who then explained that the sound be had heard was the most heavenly music that mortal ears listen to – the sound of a beautiful young white girl offering her grateful and innocent prayers to the Almighty for all the blessings He had vouchsafed to her. "I wish you had been there to listen," he said to the Prayer-wallah, who came on duty at the same time and later told me the story, "it would have taught you something in your own line."

This was the only two hours' sentry-go that I did. The same morning another man arrived to take my place, and I was ordered to report back to the Battalion. On the afternoon of the following day the Prince was giving a garden party inside the grounds of Sikundra Taj, which was six miles from Agra. I had been taken off the guard to accompany the junior Signalling-Sergeant and a corporal to this Taj, in order to do a little signalling work with the Fort on the following day. We travelled there by gharri, which was a kind of cab drawn by two horses, taking with us a heliograph and lamp. All we had to do was to get into communication with the signal station at the Fort and send them a prearranged signal, the moment the garden party broke up and the Prince left the grounds, so that the gunners there could fire a Royal Salute. We had a warrant enabling us later to travel back to Agra Station in the special saloon-carriage which brought to the garden party all notabilities who did not have carriages, and which would also take them back. There was no railway station at Sikundra but a short distance from this Taj ran the main railway line, where a temporary platform had been built for the convenience of the distinguished guests.

Not one of us signallers had visited Sikundra Taj before. It was the tomb of the mighty Akbar who conquered the whole of Northern India and was the most enlightened ruler that India ever had. I later got well acquainted with the place. Inside the grounds, opposite the lofty arched entrance, is an underground passage of fifteen or twenty paces which leads to Akbar's tomb. In the roof overhead is a slit so cleverly cut that for the most part of the day it conducts a ray of sunlight on to the head of the tomb. Inside the spacious grounds are the graves of some of Akbar's ministers of state; and last, but not least, the graves of nine British soldiers who died there of cholera many years before I was born. We arrived at the Taj about an hour before the garden party began. A high staff-officer was at the entrance taking tickets. He knew all about us and, when we told him that we were going to fix our helio and lamp on the flat roof on which the minarets were raised, he told us that if we had not been up there before we had better have a guide, because the steps leading up to it were very puzzling. He called a native to lead us up to the flat roof, and told him and another native to be ready to lead us down again at the conclusion of the party. After going up one flight of steps we came to a group of four doorways, each with a flight of steps leading up or down to some other part of the building; we took one flight which led upwards and this brought us to another set of doorways, each with its flight of stairs. At last we reached the roof, which was a decent height from the ground. The marble minarets built on it were very much higher than those of the Taj Mahal. We fixed our helio and lamp and were soon in communication with the Fort. We had brought the lamp in case darkness set in before the garden party ended. The atmosphere was very clear, with the sun shining strongly. In the afternoon a thick blue haze spread over everything and we could not see a mile ahead of us, but as long as the sun was shining the signals of a heliograph could easily be made out through a haze twice as thick this one.

We passed the time away in watching the guests arrive. The dress of the native Princes contrasted oddly with the frock-coats and top-hats of the white Big Pots, who must have been sweating a bit in that strong sun. One prince had a diamond in his turban which made our mouths water. Corporal said that if he owned it he would immediately sell it and purchase a brewery for his own private consumption. After the guests had been presented to the Royal couple, they collected in groups, walking up and down the grounds. They all looked as solemn as owls and a few stiff drinks would have done them the world of good. If there was a refreshment-bar inside the grounds we could not see it, even from our excellent vantage-point. The Sergeant remarked that if ever he climbed the social ladder and was invited to a party like this, he would get three parts drunk before presenting himself, and would make sure of being perfectly drunk before leaving, by stuffing a quart bottle of whiskey into the tail of his frock-coat.

Late in the afternoon the Royal couple, who had been visiting the places of interest in the grounds, arrived among us on the top of the roof. An elderly antiquarian was with them, giving the history of some of the principal buildings thereabouts. I could have listened to him for a month, and the Sergeant said later that the old professor knew a damned sight more about the place than the Moguls ever did themselves. As the Prince and Princess approached us we stood rigidly at attention, but the Prince immediately told us to stand easy. The Princess was interested in the heliograph, which the Prince, who had been an officer in the Navy, had a good knowledge of, to judge by the way he answered her questions. A little demonstration was then given. The Sergeant fixed the sun spot and instructed the Princess to press down on the sending key which exposed the light. Back through the blue haze came at once an answering flash from the heliograph at the Fort, which I think rather surprised her. Before leaving the roof they both graciously thanked the Sergeant, who during the years that I knew him was never tired of relating the story of how he had been specially called upon to give the future Queen of England a lesson in the art of sending on a heliograph. At six o'clock the garden party broke up and the Royal couple came out through the entrance to the grounds, we sent the prearranged signal to the Fort. Just as the Royal carriage moved off the first round of the Salute was fired.

Five minutes later the sun had sunk below the horizon and the shadows of the short Indian twilight had begun to fall. Guests who had carriages were moving off and guests who had none were making their way to the temporary railway platform where the engine and saloon-carriage were waiting for them. We had been told that this train was moving off promptly at half-past six and that if we missed it we would not be able to hire a gharri at the village of Sikundra. A six mile tramp back to Barracks was nothing, if it had not been that we had our signalling equipment to carry; and the B.B. lamp with its box was a weighty proposition. We packed up our equipment and the Sergeant shouted down to the grounds below for one of the native attendants at the Taj to come up and guide us down. He shouted several times but there was no answer, so we decided to find the way down ourselves. The Sergeant led the way, saying that he had wasted some precious minutes in shouting for a guide: he could easily find his way down, as he had taken particular notice of the turns as we came up. At the bottom of the first flight of steps the light was bad, but the Sergeant without any hesitation started down a second flight – which brought us to a small tower-room with no steps of any description leading out of it. The Sergeant began to swear, which caused the Corporal and me to burst out laughing, which made him swear a bit more. Then he said: "Yes, my beauties, it will be no laughing matter if we lose that damned train. If we do miss it, and I can catch that damned soor who guided us up here, I shall bury him under one of these blasted steps, and tell the old professor to make mention of the crime in all his future lectures."

We retraced our steps to the landing above and decided get up on the roof again to shout once more for one of the attendants. No doubt we could have found our way safely to the grounds below, given plenty of time, but there were several possible mistakes to make and it was now getting quite dark. Back on the roof, the three of us sent up a shout for one of the attendants, loud enough to have awakened old Akbar in his tomb. We were immediately answered by a native below, who shouted back: "Me coming, Sahibs." He was the same native who had guided us up to the roof, and he said that he had been on the other side of the grounds doing some work when the Sergeant had first shouted. He now demanded buckshee before leading us down, which he got by the Sergeant seizing hold of him by the throat and threatening to strangle him and throw his body over the edge of the roof. The Sergeant then spun him around and, gripping the neck of his native tunic, kicked him all the way down the first flight of the steps. By the time we arrived safely in the grounds below the guide was howling for mercy, and the Sergeant gave him a final kick that sent him hobbling and shrieking across the grounds. No doubt he and the other native who had been warned to guide us down knew that we were travelling back by train and had decided between them that one of them would appear late on the spot to guide us down and demand buckshee, which he would share with the other. The pity of it was that one of them received all the buckshee, which no doubt he would have gladly shared with his chum.

We were only just in time: as we tumbled into the carriage with our signalling equipment the train moved off. The Corporal, who owned a Crown and Anchor board, much regretted that he had not brought it with him; he was sure that the Big Pots and especially the grand dames, who looked a sporting crowd, would have been delighted to have a little flutter on it to while away the tedium of our ride back to Agra. We easily got a gharri outside Agra station to take us back to Barracks, and when we arrived there the Sergeant was feeling so highly bucked that he went and brought a bottle of whiskey and two bottles of soda from the Sergeants' Mess, so that we could all celebrate the Royal garden party at which we had been present, conversing with Her Royal Highness and mixing with the guests.

The following day there was a grand procession, which was headed by the Royal Couple. All the native Princes who had arrived at Agra to pay homage to the Prince turned out in their pomp and splendour. For over a week we had signalers at the railway station to semaphore the names of the Princes, as soon as they arrived, to the gunners at the Fort; so that they could give them the salute of guns to which each was entitled. Some were entitled to larger salutes than others. Those who were only entitled to a small salute would have willingly paid lakhs of rupees for another few rounds to be added. Each Prince had done his best to display in this procession as much wealth as he possibly could. The diamonds and precious stones that some of them were wearing dazzled my eyes as they drove slowly by in their open carriages. Their horses were magnificent and many of the carriages had gold and silver mountings. But the turnout that I admired most was a conveyance drawn by four of the cleanest racing-camels I have ever seen. Standing up in it was a Prince simply dressed in white silk with a drawn sword in one hand and a shield in the other. I do not know who he was, but he was the only one who looked a real native Prince: the others looked like stuffed and gilded peacocks. From the surrounding villages many thousands of natives had turned out to watch the procession. They were in ecstasies of delight at this display of wealth and pomp, yet they themselves were only living by the skin of their teeth in dirt and extreme poverty. A few months later thousands of them died from plague, which was raging in the neighbourhood, and the survivors died in further thousands from the great famine that followed in the wake of the plague.

The Prince dined at our Officers' Mess on Saturday evening, which was also a proud day for Wales because it was the Saturday that the famous "All Black" New Zealand rugger team was beaten, for the first and only time, at Cardiff. The next morning the Royal couple attended the Protestant church, and the following day they left Agra. Our Royal family was always popular with the rank and file. The Prince was voted a trump, and I heard many a hardened sinner say over his beer at the Canteen that the Princess was a woman that any man would be proud to have either as a wife or a mother.

NATIVE SERVANTS
AND PROSTITUTES

With the exception of two very old ones, the Agra Barrack bungalows, which each held a company of men, were so constructed that a man could walk from his room to the lavatories and latrines sheltered from the rays of the sun. Outside each bungalow was a large, deep, bricked well. One of the two old bungalows was a double-decker. On the lower deck were the library, the billiard-room, the Army Temperance room, and the coffee-bar. On the upper deck, in rooms partitioned off, lived the Signallers, the Drummers, the Regimental Police and the military telegraphists. In between the rooms were small landings where the cleaning boys did their work and where the punkah-wallahs pulled the punkahs during the summer. The rooms were so built that there was a door by every bed; two beds were close together, then came a wide door and so on all around the room. Over each two beds was a punkah-pole with a thick piece of cloth attached to it measuring about six feet by three and hanging at a height of about two or three feet above the beds. In the centre of the room was the main punkah-pole to which the punkah-wallah attached the rope which set all the punkahs in motion when he pulled it from outside. The punkahs began working on the 29th March every year and finished on the 29th October. During the whole of this period the punkahs were in constant motion, day and night; swaying to and fro over the beds they made a gentle breeze which was most welcome in a temperature that sometimes stood at 121 in the shade.

There were two swimming-baths in the Barracks: a decent-sized one at the back of the double-deck bungalow, and one behind the gymnasium. In the summer the companies paraded by sections for dips in these baths; though any man could bathe there when he liked. Speaking of these baths: it is interesting and necessary, if one writes of events and places after a lapse of thirty years, to get an opinion on what one has written from people who were present, and so check up the details. There are still a few of the old hands knocking about in Blaina and the neighbouring villages, and to my surprise they assured me that in the account I first gave of the Agra barracks I had made a great mistake. I had supplied every bungalow in the place with a swimming-pool, instead of there only being two for the whole Battalion. I suppose the mistake arose through the principal one being attached to the double-decker bungalow, so that we Signallers used it at all hours of the day, and my memory got twisted so that I imagined every bungalow had the same convenience. My chums also reminded me of something that I should certainly not have forgotten to mention, which was the lawn close to the swimming-pool, where twice a week, in the evening, the Band rendered popular tunes and selections from operas and musical comedies. These concerts meant a lot to us: even the most hard-bitten of us were passionately fond of music and the tunes revived memories of happy days and nights in England.

For the first three or four months of the summer, from nine in the morning to six in the evening, all the doors in the bungalows were fastened back and a kind of fibre-and-straw door, called a tatty, was fitted into the doorways on the hot side of the bungalows. Outside each tatty was a large earthenware bowl, and a native called a bhisti-wallah, who carried a goat skin of water, was continually travelling to and fro from the bungalow to the wash-house and refilling these bowls. Other natives, called tatty-wallahs, stayed outside the tatties, and it was their job to dip a small tin into the bowl and throw water over the tatties. This also helped to cool the room somewhat. The tatties were taken down in the evening and put up again in the morning. The punkah-wallahs worked in reliefs of eight hours a day. They, the tatty-wallahs and the bhisti-wallahs were paid five rupees a month, which in English money was a daily wage of a fraction under twopence three-farthings.

Native sweepers cleaned the bungalows and the verandahs. They and the natives who cleaned the latrines were paid four rupees a month. The native cooks received a monthly wage of five rupees, but we always had our own cooks supervising their work. The native barber, or nappy, attached to the Battalion employed a couple of assistants: for two annas a week they would shave a man twice a day if he needed it. They would come around the tents or bungalows very early in the morning with a hurricane-lamp and shave the men before reveille sounded. The head nappy had been known to shave men who were heavy sleepers without waking them up.

All the washing was done by the dhobis, as they were called, for which each man had twelve annas a month deducted out of his pay. A man could send fifty pieces to the wash every week if he had them, and he would have to pay no more than if he only sent five. In addition to shirts, towels, singlets and socks the dhobis also had to wash the khaki and white suits of the men. They generally did their washing in a stream where there were large flat rocks on which to beat the clothes they were washing. Although they used no soap, the clothes were wonderfully clean when they brought them back, but any man who had given the dhobi a punch in the stomach would be repaid next washing-day by having his washing beaten on the rocks with such vehemence (accompanied by a stream of curses) that it would come back full of holes. The dhobi would try to explain away these holes by saying that the pieces were so old and rotten that the holes appeared in them the moment he gently dipped them in the water. Every section of men had its own cleaning-boys who cleaned the boots and equipment, also the buttons and badges. The only things they did not clean were our rifles and bayonets, which we never allowed them to touch. The cleaning-boys had to find all their cleaning requisites and run errands, for which they were paid four annas a week by each man who employed them.

There were not many men who did not take advantage of having such cheap valets. These natives managed to exist very well on their low wages. Clothes did not cost them much; a rag to cover their loins and a turban on their heads was all that they needed. Their food did not cost much either: they could manage on two meals a day. Two pice-worth of flour mixed with water and kneaded into thin flat cakes, called chappaties, was their morning meal, and their evening meal was curry and rice. Many of them were rearing families on their wages. All the natives mentioned above, with the exception of the sweepers and latrine-wallahs, belonged to some caste or other: they would never eat any of the food that we left. But the sweepers and latrine-wallahs were of such low caste (or no caste at all) that they did not care in the least what they ate. After we had finished our dinner our sweeper would appear on the scene with a very large tin, in which he would pack everything that we had left. He would take it down to his bosom pal, the latrine-wallah. The two of them generally sat behind the latrine to eat our leavings. It was a nice shady spot and the smell that hung around did not seem to blunt their appetites in the least: no matter how full the tin was, they never failed to empty it. They did not believe in a routine of two meals a day: they would eat whenever they got the chance. As they lived so cheaply they could afford to indulge in a little gamble now and then. They were passionately fond of a little gamble. Near the latrine a school of four or five of them would collect together and play their favourite game of dice. Three small dice numbered from one to six were used, and each man in his turn shook them in his hand before throwing them on the ground. The man who threw the highest won the pool. The stakes were never high. The lowest Indian coin was called a pie, worth one-fourth of a pice, and one pie was the stake generally agreed upon. Before each game began each man threw his pie into the pool. I sometimes watched them play, the stake rising to two pie and three and finally to a whole pice, and then they became so excited that they would have a difficult job to hold the dice in their hands, they were trembling so. No gamblers at Monte Carlo playing with thousand-franc chips ever got more excited than what these natives did when they reached the stage of staking a whole pice on a single throw.

Our two cleaning-boys were father and son. Whether or not the father paid his son, who was about fifteen years of age, out of the wages he received on behalf of the both of them I can't remember. Most of us paid the cleaning-boys regularly once a week, a few were none too punctual, and Crickett never paid at all, unless it was with a stream of the crab-bat. Private Crickett was a brilliant signaller but also a bit of a hard case. Whenever he got very drunk he used to mess his clothes, which the cleaning-boys always had to wash and dry. One day the elder boy came to me and said: "Oh, Dick Sahib, Crickett Sahib plenty bad man. Some mornings soon as me come on verandah, he shout 'Boy come here, you black pig.' Me run to him and he throw shirt, pants, trousers in my face. He say 'Clean them, you black bastard, or I'll knock hell out of you.' Shirt, pants, trousers, plenty muck: me plenty clean. Crickett Sahib never pay me. Not pay me one pice all many twelve months. Plenty bad man, plenty hard man, plenty fierce man, Crickett Sahib."

I knew the boy was speaking the truth, but he was not the only native that Crickett did not pay. Although the nappy shaved him every morning and cut his hair occasionally he never paid him a pice either, though he sometimes promised that he would, some day. He and several more of us were having tasty breakfasts and special dinners and little luxuries for our tea, for which we each had agreed to pay the native cook a rupee a week. Crickett paid for the first two weeks but was never known to pay afterwards. In spite of this, the two cleaning-boys would spend more time in cleaning his things than any other man's, and he was always the first man whom the nappy shaved, and always the first man to have his food served by the cook. I asked him one day, wasn't he afraid of having a special dinner served up to him, with powdered glass in it, from which he would die in awful agony after eating only a mouthful or two. No, he said, he wasn't. He did not believe in the yarns that had been handed down from bygone days of how the native cooks used this method of doing-in the men who had given them a hiding or who had failed to pay them what was owed. These yarns could be heard in every battalion in India and Old-Soldier Carr used to swear that he had lost his best chum in this way; on the day his chum was admitted to hospital the native cook vanished and he was never seen again. Crickett said that his plan was have them all hope that one day he would declare a dividend, and that when he did the men who had been most attentive to him would be granted a preference over the others. But he never paid one of them. When he left the Battalion, time-expired, I expect a good many prayers went up to the gods of India that the troop-ship that was taking him back to Blighty would sink in the deepest part of the ocean.

At Meerut, Kilana and Agra were brothels reserved for the use of white troops. The one at Agra was in the Suddar Bazaar, which was about three-quarters of a mile from the Barracks. In this brothel, or Rag as it was called by the troops, were between thirty and forty native girls whose ages ranged from twelve to thirty. This number was considered sufficient for the fifteen hundred white troops that garrisoned Agra. Each girl lived in a separate shack of her own, which was made of plaster and mud with a hard-baked mud floor. The only furniture was a native rope-bed with no bed-clothes, a large earthenware vessel for holding water, and a small wash-hand bowl. Our Regimental police relieved one another in patrolling the small street which the Rag was in. Natives who passed through this street were not allowed to stop and talk to the girls; if any one of them did, the policeman would give him such a thrashing with his stick that he would remember it for a long time. The Rag was opened from twelve noon to eleven at night, and for the whole of that time the girls who were not engaged would stand outside their shacks

148

soliciting at the top of their voices and saying how scientific they were at their profession. There were always small native boys between the age of six and nine knocking about this street. They earned an occasional pice by running errands for the girls and many a pice, during the summer, by acting as punkah-boys. Some of the shacks had covered-in passages, about six feet in length, leading to the small door of the room. In these shacks some of the occupants had erected a small punkah over the bed for use during the summer, with a rope, through a hole in the door, which was pulled from the outside. The girls who possessed punkahs considered themselves very up-to-date. They did not charge anything extra for them but neither did they pay to have them pulled when a man was in the room with them. If a man wanted a gentle breeze during his short visit he would engage the punkah-boy outside the door, who charged one pice for the job. They were wicked little devils and at the early age of eight they knew more about sexual matters than the majority of grown-up men. Standing outside the door with the punkah rope in their hands, they would cry as soon as a man stepped foot inside the passage: "Sahib, Sahib, you want it the punkah pulled while you jiggy-jig, me very clever, me look through hole in door while you jiggy and still pull punkah, me pull slow, me pull fast, me mallam which wanted, me very good punkah-wallah, Sahib."

Everything possible was done to prevent venereal disease. Each girl had a couple of towels, vaseline, Condy's fluid and soap; they were examined two or three times a week by one of the hospital-doctors, who fined them a rupee if they were short of any of the above requisites. If he found that any one of them had the disease he had her removed to the native lock-hospital. There was also a small lavatory erected in the street, which had a supply of hot water; it was for the use of any man who was not satisfied with the washing he had done in the girl's room. There was always a number of men in hospital with venereal but it was very rarely that they contracted it in the Rag. It was generally caught from half-caste prostitutes or native girls outside. If a

149

man hired a gharri to go for a ride somewhere the driver would immediately say: "Sahib, you want nice Bibi, me drive you to bungalow of nice half-caste, plenty clean, plenty cheap, only charge one rupee, Sahib." In the evening a man could not take a walk anywhere at a distance from the Barracks, without hearing the familiar cry of "jiggy-jig, Sahib." Very small boys did the soliciting for these native girls, who being in the last stages of the dreaded disease and rotten inside and out, only appeared after dark. These were the sand-rats and it was a horrible form of suicide to go with them. I always remembered the old time-expired man's advice at Deolalie and never forgot to pass it on to young soldiers who were fresh to the country.

For the benefit of our health, because it was said that to abstain was unhealthy in a hot climate, the Prayer-wallah and I occasionally visited the Rag. I took every precaution, for which I am truly thankful, and most of the other Signallers the same; but no man that ever breathed ever took more cautions than what the Prayer-wallah did. In addition to septics and prophylactics he possessed a powerful glass with which he would minutely examine the choice with the professional manner of an old family doctor. He could never understand why the Indian Government did not issue every man in India with one the same. He said that it was of far more benefit to him than his tooth-brush, about which so much was spoken in the addresses on Personal Cleanliness that we were sometimes given. Whenever he appeared in the Rag-street the girls whom he had been with before would shout: "Hullo, my Spy-glass wallah!" But in spite of the generally accepted notion that it was impossible for a healthy young man in India to keep away from women, I knew a number who did. I always admired these men and still correspond with a couple of them. One of them is a railway detective-sergeant in Lancashire and the other who rose to the rank of captain during the late War is doing well in business in the Midlands.

The Suddar Bazaar was on the right of the main road leading to Agra. Opposite the centre of it, on the left of the road, and standing on its own, was the Protestant Church. It was possible on a Sunday evening to stand in the road outside the church and hear, on one side, the parson with his monotonous clerical voice preaching about the spiritual joys of life, and on the other side the shrill and equally monotonous cries of the girls in the brothel advertising its material joys.

Between the Suddar Bazaar and a larger bazaar was a small group of shops that manufactured and sold imitation and pure white marble models of the Taj Mahal; only the wealthy could purchase the expensive sort, which were exquisitely made. Here was also manufactured and sold a curious kind of stone which could be bent like a piece of rubber. During the first month at Agra we were allowed in the large bazaar; then it was rightly put out of bounds. It was an evil-smelling place and full of temptations for a young soldier. The city of Agra was always out of bounds for British troops, but once a year the Battalion paraded in fighting order and marched through it with fixed bayonets. The only explanation that I could get for this ceremony was that two-thirds of the population never moved outside the city, so the resident battalion had to march through to remind them that the British Raj was still in force over India.

In India native parents invariably pay more attention to their boys than to their girls; the only exception that I know of was a village about seventeen miles from Agra where the girls came in for all the attention. The soil around this village was not very productive and the people had a difficult job to exist, so parents with daughters denied themselves a little food in order that their girls should have well-nourished bodies by the time they reached the age of ten or twelve. They were then taken by their fathers to a certain market in Agra and sold like prize-cattle to the highest bidders. The plumper the girls were, the more money they would fetch. Their purchasers were sometimes rich old men, who used them for their private pleasure, but more often they were the managers of houses of ill-fame in the City, where the girls would soon be installed. Every pice that they earned would be taken by

151

their masters, who would only provide them with food and a few rags to cover their bodies. Even if it had been possible for them to escape they had nowhere to go to. It was only a question of time before they would be in the last stages of disease and of no further use to their masters, who would then drive them out of the house, so that they became sand-rats until they died. This village was practically exterminated by the plague and famine that I mentioned at the close of the chapter about the Royal visit. It was a damned pity that it ever existed; although girls in other parts of India were sold for the same purpose by their parents, and very few of the girls in the Rag that served for white troops had had any say in the original choice of their profession. But this much can be said for the Rag, that the girls were paid good money and were well cared for and, so far as we knew, they were their own mistresses: if they chose to hand over their earnings to a father or a brother, that was no concern of ours. None of them seemed to spend much time in lamenting their hard lot.

I might as well make a general statement here about myself and women. It is this: I have knocked about with a number of them in my time but have never married. There was only one woman I ever really cared for, and she married someone else, and it was my fault that she did. I expect that this is saying enough on the subject. But I have a lot more respect for women than what perhaps appears in this story of my young days.

SPORT AND FEVER

At Agra in the winter it was warmer during the day than what it was at Meerut, but colder during the night. All kind of sports were indulged in: boxing tournaments and soccer matches between companies were the most popular. Every morning about breakfast time natives would come around the bungalows shouting: "Live hare Wallah." Across their shoulders was a bamboo pole, from each end of which hung a large basket containing hares or wild cats, or a jackal which they had trapped alive. They were clever trappers, for it was very rarely that their animals showed any sign of injury. Strong hares, fit to run for their lives, cost six annas apiece, wild cats cost four annas apiece, and a jackal, if he was a big one, could cost as much as eight annas. Those of us who had greyhounds or other speedy dogs would buy the hares for sport, but it was difficult to keep a coursing-match private: by the time the hare was released all the dogs of the Battalion, including the greyhound of bulldog ancestry, would be on the spot and joining in the chase. This dog's owner used to say that if it ever had the luck to catch a hare it would never leave go its hold, that was a safe bet. On a wide open maidan it was possible to follow the chase right to the sky-line. The hares were given a sporting chance to regain their freedom but it was very rarely that one of them reached the sky-line before the greyhounds caught it. The wild cats were often twice as large as house cats; they were also given a chance to regain their freedom but not one of them ever did. They were a treat for the dogs that were not speedy enough to catch a hare: having no tree to run up they were usually overtaken and killed after a fifty yards' chase.

Jackals, which are like a cross between a wolf and a dog, were never given any chance: they were taken to the ball-alley, where the men played hand-ball, and there made to fight to the death against dogs. A jackal fighting for his life was a dangerous, cunning beast, so two or three fighting-dogs were always pitted against it. The men entered the ball-alley, secured the door behind them, and held their barking dogs in check until the trapper released the jackal. Ordinarily a jackal was a bit of a coward, but when he realized that there was no possible chance to escape, the walls of the alley being too high for him to jump over, he would spring to a corner and turn like a flash to face the dogs. With arched back and bared fangs he looked a frightful creature; his jaws clashed together like steel traps, and if his fangs connected with the bony part of a dog's leg, that dog would have to be destroyed. He fought like a cornered rat. But the odds were against him; in ten or twenty minutes he was either dead or shamming dead so well that he would deceive even the dogs. The first time I saw a jackal shamming dead I was completely taken in, and so were the other men present: even the dogs, who had withdrawn from him, must have thought the same. But the trapper said: "Him not dead, Sahibs, him plenty sham dead, you bring bucket of pony (water), throw over him! That make him plenty jump high, Sahibs." A bucket of water was brought and thrown over him. The effect was magical: he sprang in the air a good six feet and the dogs, who seemed to realize that they had been tricked, were at him madder than ever. This time their owners had a difficult job to make them release their holds after he seemed dead a second time. The water-test was applied again, but there was no need for it.

Boxing was compulsory. At Agra on physical training parade we did it by sections. Each section consisted of sixteen men. Gloves were served out and the order was given to extend to four paces interval, and then: "Front rank, about turn! Box!" We went at it, front rank against rear rank, hammer and tongs, for five or six minutes and then had the pleasure of watching the other sections go at one another in the same way.

In the two winters we had spent at Meerut the Prayer-wallah and I never missed any of the small race-meetings that were held there. There was no race-course at Agra, but in the winter a small race-meeting was held on the maidan about a mile from Barracks. The majority of the troops in the garrison attended it. The races were for Arab horses over a distance of four or five furlongs, and the European bookies who appeared mysteriously from somewhere or other, like vultures, did a brisk trade. The Prayer-wallah and I were one day returning silent and stony-broke from a race-meeting when he suddenly burst out: "Well, Dick, in my time I have read the Bible through from cover to cover, over and over again, but there's one thing I'll never believe, and that is that the Jews are God's chosen people. If He has any chosen people at all it is the Bookies, and the ones we laid our bets with to-day must have been His own peculiar darlings."

On the Plains it was impossible to carry out long-distance signalling except from high buildings some miles apart. Agra and its surroundings were ideal for this kind of work. I don't suppose that the old Mogul Emperors would have built their beautiful tombs and buildings at such a height if they had known that after their time the tops of them would be used by British signallers as convenient posts for fixing their heliographs and lamps. After our annual signalling test we did long-distance work, each station remaining out for a week or ten days. About twenty-three miles from Agra is the old ruined city of Futtehpur Sikri, which was built by Akbar for his own residence, but the water supply gave out and he removed his court back to the Fort. The city was soon abandoned: with the exception of some priests, and some natives employed at the Dak Bungalow (a hostel for travellers), it was now only the home of pariah dogs, jackals and such-like animals. There were still some high buildings there in a good state of preservation, including a huge sandstone mosque. At the back of the mosque was a beautiful mother-of-pearl tomb, with exquisite marble screens, which had been erected by Akbar in memory of some holy man. The tomb of this holy man was carefully looked after by the priests. They would not allow anyone to go inside it with his boots on; which was a thing not insisted upon at the Taj Mahal or any of the other tombs in this area. We had a terminal station on the roof of the sandstone mosque, a transmitting station on the roof of Sikundra Taj, and a terminal station either at the base of the large dome of the Taj Mahal or on the roof of the Palace at the Fort.

Signalling equipment and cooking utensils were taken by bullock-cart to Sikundra and Futtehpur Sikri. Sufficient men were always sent to these two stations so that a couple of them could have a day's holiday in turn. Our Signalling Officer, Mr. Venables, who was a trump, used to lend one of his sporting guns with plenty of ammunition to the station at Sikundra. I much preferred this station to the others and so did the Prayer-wallah, who possessed a revolver with a good supply of ammunition which he had brought from South Africa. There were herds of wild deer around this place, but it was not easy to get within shooting range of them: as soon as they scented human beings they were off like the wind. Occasionally one was shot with the sporting rifle, which made a welcome addition to the grub stakes. A few of the men amused themselves by shooting the jackals that were sometimes to be seen skulking about during the day, or any pariah dogs that crossed their path. But the Prayer-wallah never wasted a round from his revolver on one of these animals: he used to say that they were not worth powder and shot. Cheetahs and other animals lurked in the grounds at night, but our camp-fire, which was constantly replenished by the sentry, kept them away. Our chief trouble at Sikundra were the monkeys. They were about three feet six inches in height and wanted constant watching; they were more daring thieves even than the monkeys at Chakrata. The Prayer-wallah and I spent many happy hours exploring the ruins of these old places. Every night around the camp-fire we would indulge in a sing-song, and although we were among the tombs of kings and saints we sung the choruses of the latest music-hall ditties with as much gusto as if we had been in the Canteen.

In April 1904 service-pay was introduced, but men had to have completed two years' service before they were entitled to it. To qualify for the top rate of pay a man had to have at least a third-class certificate of education and also be a first-class shot. Men of two to five years' service who possessed both these qualifications were entitled to sixpence a day extra pay; if they had only one of them they received fourpence, unless they were only third-class shots, in which case threepence was all they got. The top rate, for a man with five years' service or over, was sevenpence a day, which could be dropped to fivepence and threepence, according to his qualifications. But a man even on the top rate of pay was dropped to threepence if he was convicted of a crime for which he received an entry on his Regimental conduct-sheet. I knew many first-class soldiers who never received the top rate of pay, merely because they did not have a third-class certificate of education. Some of them could read and write a little, but they hated the thought of going to school again and would rather be paid less than do so. I hated the idea of school myself, but I did not intend to have twopence a day less than another man, not if I could help it. I started to attend afternoon school a fortnight before an examination came off, and then sat for it. The teaching was something similar to what my cousin Evan and I had been given in Standard Three of our board-school when we did not happen to be playing truant. I managed to pass the examination and was awarded my third-class certificate.

During the winter of 1904 we lost a number of men from enteric fever and a few from abscess on the liver. A corpse was only kept twelve hours before burial and the funerals were held early in the morning or late in the afternoon. On the way to the cemetery the band played that mournful dirge, The Dead March; but on the return journey they played the airs of the latest comic songs, which helped to dispel the gloom. If a man died, only his company had to parade for his funeral; if he had pals in other companies they were permitted to attend it too, but it was very rarely that they availed themselves of the permission. Every

Thursday was a general holiday for the troops in India, but if, say, 99 Jones died on Wednesday evening or before sunrise on Thursday morning, his company would be docked of a part of their holiday by having to parade for his funeral either in the morning or in the afternoon. Many vigorous remarks would then be made by the men of 99 Jones's company. "Another bloody holiday we are done out of," one man would grumble. And another would chime in: "Yes, I always thought old Jonah had a bit of sense. But he couldn't have had such a hell of a lot, otherwise he would have waited until tomorrow to kick the bucket." "Or got it over with by Tuesday if he couldn't stick it out," another would say. And yet there wasn't a man in the company who would not have done everything in his power to alleviate any of 99 Jones's sufferings while he was alive. And while he was alive 99 Jones had, no doubt, made similar remarks when warned to parade for a funeral on a Thursday. I believe that many a dying man did his best to last out the twenty-four hours ending at sunrise on Thursday, so that his company would have a good word for him after he was gone. During my time there were none of the rank and file of the Royal Army Medical Corps stationed in India or Burmah. Each Battalion had its own trained hospital orderlies who had permanent staff jobs in the station-hospitals wherever their battalions might be sent. In addition to the medical officers and white nurses in the hospitals there were certain half-caste apothecaries whom we called Black Pots. Most of the ones I knew were very able men, who knew as much about diseases and their cure as some of the white doctors did. There were a number of native orderlies in each hospital, who had to keep the wards clean and do all the rest of the dirty work. The sweepers and other wallahs were paid the same wages as those in the barracks.

It was the custom at this time to send two parties to the Hills during the summer. The first party, which consisted of one company, with the invalids of other companies attached, spent the first half of the summer at a Hill-station. They would then return to the Plains, and the second party would take their place. The first half of the summer was the most severe. Many of the invalids who went with the first party were men who had been in hospital with fever, or men who did not look too well and had been picked out by the Medical Officer at an inspection held for that purpose. Some men managed to work it to be sent to the Hills every year with the first party, and to stay there the entire summer; these were sarcastically called Hill-parrots by the men who did not have the luck to go to the Hills at all. A year or two later the custom of one party relieving the other at the Hills was abolished; the first party remained there for the whole of the summer, but small batches of invalids were still sent to different Hill-stations as soon as they were fit to travel.

The first summer at Agra we Signallers were sent to Kilana with the first party. We travelled by train to Dehra Dun and in the first week in April 1905 we began our four days' march to Kilana. On the last day's march we took the short cut. The morning was hot and sultry, which was unusual for this time of the year in the Hills. After marching for about an hour we felt an earth-tremor which made us wobble about like old soldiers leaving the Canteen late on Christmas night. None of us had experienced an earthquake shock before but we soon realized what it was, and we all had the wind up. There was another shock a few minutes later and this time a number of men were thrown violently to the ground. The Prayer-wallah made us all windier still by saying in gloomy tones that he would not be surprised to see the ground open and swallow us up for all the grievous sins we had committed, the same as it had done to Korah, Dathan and Abiram in the Book of Exodus. Shortly after our arrival at Kilana there was a third shock. The bungalow I was in seemed to wobble a bit for a second or two, and we all ran out into the open. When we returned we found that several of the bungalows had large cracks in them, but none had collapsed. This was the last shock we experienced.

Later we heard that the earthquake had been felt over an area of hundreds of miles. Many thousands of natives lost their lives and much damage was done to buildings. A battalion of Ghurkas stationed at Dhamsala lost two or three hundred men, but so far as I remember no British soldiers were killed. Lives were lost at Chakrata, Landaur and Mussoori. At Mussoori a new church was being built and was very nearly completed, but the earthquake levelled it to the ground. Many of the natives who were working on it were killed or injured; those who had escaped unhurt would not work on it again, saying that their gods were angry with them for working on the White Sahibs' church, and therefore had sent the earthquake to destroy it. Those of the White Sahibs who happened to be religious men said that it was an unfortunate accident; and the One Above was not blamed, because the building had not yet been consecrated. The monsoon was early this year. We marched back to Dehra Dun through torrents of rain, to the accompaniment of tremendous croaking from the huge bullfrogs in the fields as we passed by.

We were back at Agra about the middle of July. There had been no rainfall at Agra, which had a very dry climate. We were here over eighteen months before a spot of rain fell. Then it came down in sheets for nine or ten hours. Another twelve months passed by without a drop falling and then it rained again for about the same time. In spite of this absence of rain there was plenty of fresh water in Barracks: the bricked wells outside the bungalows supplied all that was needed. The natives living around Agra also depended on bricked wells for their water, and these had in some places been sunk and lined by the Government. But the agricultural natives had to water their crops from their primitive unlined wells, and if these caved in or went dry, as they often did, their crops would be ruined. The failure of these wells was one of the chief causes of the famine of 1906.

Outside our bungalows were some large trees which were the home of hundreds of small parrots; but it was impossible to train one of them to talk. Several had been caught, but their trainers soon gave them up as hopeless propositions. Large Bengal parrots could be bought, however; these were the first speaking parrots introduced into Europe, I have been told. Many men believed that the best method of training them was to lower them in their cages down a deep well at the dead of night. This was supposed to frighten them so much that they would soon repeat what their owners were saying from the head of the well and perhaps soon be holloaing like hell to be pulled back up. One of the drummers in the next room to ours had bought a large parrot and was a strong believer in this method of training. He boasted that in three months' time his parrot would be talking like a poet and swearing like an old soldier. He had the parrot for a week or so in his room, training it, then one night at about half-past one a few of us who happened to be still awake heard a voice which we recognized as the drummer's saying: "Pretty Polly, pretty dear, all the way from Kashmir." The voice was not loud enough to awaken those who were asleep, but quite loud enough in the stillness of the night to keep those awake who had difficulty in dropping off. The drummer had chosen the well outside our bungalow and although we sleepless ones were amused at first, our amusement soon turned to groans, curses and prayers.

His monotonous voice got on our nerves. For ten minutes or so he kept to the Pretty Polly stuff, and then he would hurl down the well a volley of swear-words that were strong enough to bring a dead parrot back to life and curse him dead again into the bargain. For two hours he kept this up and if all our prayers had been answered during the first hour the parrot would have contracted galloping consumption, been blasted by fire from Heaven and finally sent to a watery grave in the well by the breaking of the rope. During the second hour the prayers were all for the drummer. The Prayer-wallah, who was one of the sleepless ones, was in magnificent form. If one long earnest prayer of his had been answered the drummer would have been struck as dumb as Zacharias, pecked to death by a flock of hundreds of Poll-parrots all the way from Kashmir and then pitched head first down the well with such force that only the soles of his boots would have been visible in the bed of the well below the water.

When morning came I asked an old drummer whether the parrot trainer had kept any of the men awake in his room. "Yes," he replied, "he did, the bastard, but by Christ that parrot will be a lucky bird if it is still alive and able to continue his education down the well to-night." At twelve noon the parrot trainer went to the Canteen leaving a strong and healthy bird behind him in his room; when he returned three-quarters of an hour later he found it dead in its cage from causes unknown. The old drummer, who was very sympathetic, said that the parrot had evidently died from rapid pneumonia, contracted from the damp well during the night. The parrot trainer, who was convinced that his bird had been poisoned, spent the rest of the day vowing vengeance on the murderer. But he could prove nothing against anyone. The Prayer-wallah, who had been in deep conversation with the old drummer during the morning, told me that the One Above moved in a mysterious way His wonders to perform, and that if a man only prayed earnestly enough his prayers were bound to be answered. This method of training parrots may have worked in a well far removed from the bungalows but it was never proved to be efficacious at Agra Barracks. The only other parrot that began its training in the well outside our bungalow also died mysteriously after one lesson.

On the third night after returning from the Hills I felt cold and shivery. The night was warm and I could not make out what was the matter with me. The next morning I had a splitting pain over the eyes, which was so intense that I found it difficult to open them more than a slit. Instead of going on parade I went sick and staggered, rather than walked, to the hospital, which was just across the road from our bungalow.

I was admitted to the hospital with a temperature of 104.8 and taken to the special ward, reserved for men who were dangerously ill and a few who were recovering from heat apoplexy. There were two empty cots in the ward, which one of the heat-apoplexy cases cheerfully told me had been occupied the night before by two men who had died just before dawn. When one of the orderlies told me which cot I would occupy, and then went to the stores to order one of the natives to bring along the bedclothes, the cheerful one said: "Well, Dick – that's your name, isn't it? – you're occupying the unluckiest cot in the ward. Since I've been here I have seen six poor devils wheeled away from it to the mortuary, and that's four more than any other cot in the ward can claim to have done for. I only hope you'll be luckier than what they were."

I felt too groggy to reply, but another heat-apoplexy case shouted at him: "You miserable old bastard, can't you keep your blasted trap closed? You try to put the wind up every man that comes in the ward. It's a hell of a pity you weren't wheeled off to the dead-house yourself, the very same day you were admitted here."

The two of them now engaged in a battle of swear-words, picturing what they were going to do to each other when they were fit again. Two-thirds of the patients in the ward were too ill to take any notice of them, and they only ceased their battle when one of the white Sisters entered the ward. Shortly after I was in bed the Black Pot visited me. He pricked the top of my middle finger (I don't remember of which hand) with a lancet and gathered the spots of blood between two small slides of glass, which he took away with him. This was the test to see if I had ordinary fever or malaria.

That afternoon I felt much the same as in the morning, but as the evening drew on I grew worse. At some time during the night I woke up to find the bedclothes stripped off me and the two orderlies bathing my naked body with sponges, which they were dipping in a bowl of iced water on the table by the side of my bed. For a moment I thought that they were giving me a bath, which would save them the trouble of doing it in the morning. I said: "This is a nice time to give a chap a bath. Couldn't you wait until the morning?"

"How do you feel now?" one of them asked.

"All right. I have no pain at all," I told him.

One of them put the thermometer under my arm, and then it struck me that I had been delirious. Before they took the thermometer out again I returned to my delirium, but did not know of this until they told me the following morning. When I came to myself the second time they were still sponging me in an endeavour to get my temperature down. At eight o'clock in the morning my temperature was down to exactly 100, but looking at my fever-chart, which showed me that I was a malarial case, I found that it had risen during the night to 106.8. The next three nights were repetitions of the first, but by eight o'clock in the morning I was either normal or only one or two points above normal. The evenings were the worst. Then day by day my temperature gradually dropped until at the end of a fortnight it was safely down at normal again. All this time I had been living on milk and liberal doses of quinine, which had also been injected into me. I was now made to go another nine days without a temperature before being put on diet. These days went by very slowly and when I did get my diet it was not a liberal one at first. As soon as they put me on full diet I began to eat like a horse, and one day I decided to ask the ward-doctor, when he came on his morning round, for a pint of Canteen beer to drink with my dinner. Men on diet who were allowed beer with their dinner had a choice between a half-pint bottle of Indian Pale Ale and a pint of Canteen beer. They invariably chose to have the Canteen beer: it was double the quantity of the other, and the Canteen-Sergeant would hesitate to water it for fear the doctors might analyze it. I chose the wrong morning for my request. The ward-doctor was a little liverish and said that in a week or ten days he might grant my request. This meant that he would not grant it at all, because by that time I would have been transferred to the convalescent ward. After he had gone, one of the orderlies said: "Don't you worry, Dick, there's more ways than one of getting a pint. You needn't depend on the ward-doctor for it. If you carry out my instructions you'll be in God's pocket. Tomorrow morning the old Colonel is visiting the ward. Maybe you've noticed that he generally comes in through the door by your bed. As soon as he enters, rise up in a sitting posture, with that fly-flapper in your hand, and bring it down with a hearty smack on the bed. It doesn't matter whether there's a fly there or not. After that, make a pretence of picking up the dead fly and dropping it in the spittoon by the side of your bed. The old chap is sure to compliment you on this, and now comes your opportunity for asking him for the pint. I'll bet you two to one he'll grant it."

Now, it was a common belief among the troops that as soon as an Army doctor rose above the rank of captain he became a little balmy and the higher he rose the more balmy he became. This Colonel of the Indian Medical Service, in command of the Hospital, was a case in point. He was a kind old chap at heart, but there were times when he used to act as if he were not only up the pole but dancing a hornpipe at the top. The two things he had on his brain were Delhi Fort and flies. I did not wonder at him having Delhi Fort on the brain. Half the men of the last two companies who had done duty there had been in hospital with acute malarial fever, some cases proving fatal. He used to declare, in various degrees of righteous wrath, that he would like to borrow dynamite and blow the place to little bits. As for flies, he made them responsible for every disease known to medical science from the common cold to bubonic plague. Each man in hospital had a fly-flapper issued to him; it had a handle about eighteen inches long, encased in leather, with a round soft piece of leather at the end of it. They were excellent weapons for killing flies, but our main trouble at this time of the year was not flies but mosquitos.

The following morning the Colonel, with all the hospital staff in attendance, visited the ward. He came through the door by my bed. I faithfully carried out the orderly's instructions. As I picked up the imaginary fly and bent over the side of the bed to drop it into the spittoon, the Colonel patted me on the back. "Good man, good man!" he said. "If every soldier was like you there would be less disease in this station." He then picked up my diet-board and asked me how I was getting on. I replied that I was getting on very well, but that I thought I would recover my strength and appetite much quicker if I was allowed a pint of beer with my dinner. "I am sure you would," he graciously replied, and marked on the board: "One pint of Canteen beer." At this time of the year the beer was lukewarm and undrinkable at dinner, but we would get the orderlies to put the bottles in the ice chest: by the evening the beer was all that beer should be.

A week later I was marked up and transferred to the convalescent ward. There were twenty-four of us convalescents. One day the Colonel paid us a visit. As he entered the ward he caught sight of one solitary fly slowly circling around the four-sided lantern suspended from the ceiling. It nearly gave him a heart-attack. The way he raved and stormed, anybody would have mistaken the fly for a man-eating Bengal tiger that had entered the ward and quickly devoured half the patients that were in it. He blamed everybody for the fly's presence, from the native ward-boys up to the hospital staff who were with him. Just as he had delivered his piece, the fly made matters worse by flying around his head. He now seemed to go stone mad, dancing like a Red Indian on the war-path. He seized hold of the fly-flapper by the bed of the man nearest to him and made a terrific swipe at the fly. He missed his objective. We were all standing smartly at attention at the foot of our beds and most of us had a difficult job not to laugh. "Every man in this ward get hold of his fly-flapper," he shouted. He then made twelve of us stand at one end of the ward, and twelve at the other. When we were in position he told the Punkah-wallah to cease pulling for the time being. Then he roared: "Now, men, I want you to swarm down on that fly like a troop of cavalry and EXTERMINATE it." Any strangers who entered at this moment would have sworn that it was a mental ward and that the Colonel, who was directing operations, was the most dangerous case of the lot. Some threw their flappers at the fly, others made tremendous swipes at it; we dodged around the beds and jumped over them, but the fly, who had quickened his pace somewhat, seemed to be enjoying himself. He refused to settle. After a few minutes most of us were too exhausted to rise our flappers the height of our shoulders to make a swipe at it. To add insult to injury, the fly, after a few easy circles around its favourite lantern, sailed majestically over the Colonel's head and out through the open window to safety. When he left our ward the Colonel was madder than ever, and for the rest of the day, the orderlies told us, it was impossible for anyone to approach him without feeling the lash of his tongue. That evening half of the men in the ward were back in bed with temperatures ranging from 99 to 102, the exertion having been too much for them.

MURDERERS, HALF-CASTES, BUN-PUNCHERS

When we recovered, the whole twenty-four of us were invalided to the Hills. After leaving the train at Dehra Dun we proceeded by tonga to Kilana, the journey being made in six or seven hours. Tongas, which were used as carts for the mail, were like small covered-wagons with two galloping ponies in the shafts. The native drivers drove them at a good speed and every fifteen miles there were stables where an exchange of ponies was made. These tongas would only carry four men, so a little fleet of them had to make the journey.

After a few weeks at Kilana, where there were details from several units gathered, mostly men recovering from fever, I was fit enough to return to duty. As there were none of our signallers there I was posted to one of the companies for ordinary duty, which was quite a change from signalling work. One evening, shortly after nine, a few of us were returning from the Canteen when we met a party of men carrying their rifles and muttering excitedly together. I said, for a joke: "Going cheetah-shooting?"

"No," one of them answered. "We're out man-hunting. It's that fellow Bern. He's roaming this hill somewhere with a loaded rifle and a few packages of ammunition. He's gone mad. He's killed a man already. My advice is to keep your eyes well skinned on the way back to your bungalows." Then the order came for all the troops on the hill to turn out with rifle and bayonet in search for Bern. Ammunition was served out to the armed pickets.

Bern, who did not belong to our regiment, had nearly completed five years' service. For nine months he had been mucking-in with a youngster who had only arrived in the country the previous winter. Both of them were fond of a gamble, but while Bern had been reasonably lucky the youngster had lost his weekly pay as regularly as clockwork. Bern had looked after the youngster like an elder brother and shared every penny he had with him. On this day, which happened to be pay-day, the both of them were in the bungalow all the afternoon where a game of Brag was in progress. It must have been Bern's unlucky day. He drew bad cards and was soon stony-broke. He had to retire from the card-school and stand in the background as a spectator, but to his pleasure and surprise his young chum for the first time for many months had a run of luck. His cards started good and grew better and better. Finally when the school had dwindled down to three he broke the other two by holding a pryle of Jacks against their royal flushes. The whole of the pay of the bungalow lay heaped on the table before him.

Bern thumped him on the back: "Well, Kid, your luck has turned at last, and not a day too soon. I got cleared out in the first five minutes. This means a jollification at the Canteen for us every night for a week. We'll live like fighting-cocks, eh?" The youngster looked up in a surly sort of way and said slowly: "You're not going to muck-in with a damned pice of my winnings. *Kooch nay*, do you understand? I'm fed up to the neck with mucking-in with you. This is where we part brass-rags."

The men who had been skinned burst out laughing, thinking that the youngster was playing a game on his chum, but Bern knew different. He flopped on his bed as if he had been struck and the youngster walked out of the room. When the men realized the the youngster had been serious they wondered why Bern had not given him a hiding. One of them said: "Hey, Bern, you can't allow that young soor to treat you in this manner. Why don't you give him a damned good belting? He's been milking you of your winnings for months now." Bern made no reply. He lay on his bed in silence for hours.

The Corporal in charge of the room and another man who was sitting on his bed knitting a blow-belt – being on the tact – were now the only persons in the room besides Bern. The others had moved off to the Canteen in the hopes of getting a drink out of the youngster. Bern suddenly jumped up off his bed and went out. At nine o'clock, which was half an hour later, he returned. He pulled a packet of ammunition out of his pocket and began to charge his magazine.

The Corporal, who was studying for promotion, was busy reading a book on Military Law. He did not notice that anything unusual was taking place, until the knitter, whose bed was next to his, leant over and nudged him. The Corporal followed his glance and saw what Bern was up to. He sprang up. "You there, Bern, why are you loading that bloody rifle? And where the hell did you get that ammunition?"

By this time Bern had charged his magazine and closed his bolt, with one round in the breech. These small bungalows at Kilana had only two doors, one at the back and the other at the front, and he was now standing opposite the front one. He said: "I went to the store-room and pinched three packages. One of them's in here" – he tapped the magazine – "and when the dirty young rat I've been mucking-in with all these months shows himself in the doorway, I'm going to fill his body full of lead. A man like that don't deserve to live. And what's more, I solemnly swear to the both of you that if you attempt to leave this room to warn him I'll shoot you dead as well. My chum is generally first up from the Canteen, so there needn't be too much fuss. But whether he's first or last, I'm going to get him." The Corporal was a brave man. He said: "Yes, by God, you seem to have picked a first-class rotter for a chum. But don't be a damned fool, Bern. There's other ways of settling an affair like this than murder. Here, lad, hand me that bundook of yours." He walked across the room towards Bern.

Bern now threw his rifle to the ready and shouted in a strangled sort of voice: "If you advance another step, Corporal, you're going to stop one."

The Corporal made a spring to wrest away the rifle, but, as he did so, Bern fired. The Corporal dropped to the ground. Bern gazed at him like a man just coming out of a trance and then rushed out of the room, taking his loaded rifle with him.

The man who had been knitting now hurried to the next bungalow and sent a man whom he found there running for the doctor. Then he returned to his own room with a group of other men, and between them they did what they could for the Corporal, who had been shot through the liver; but he was dead before the doctor arrived. The youngster was the first to arrive back from the Canteen, as Bern had expected. It was decided that it was safer for him to go out with the search-party than remain in the room, seeing that it was possible that Bern might return and try to get him. It is no joke looking for an armed madman in the dark among trees. Everybody was greatly relieved, especially the youngster, when the news was passed around, an hour later, that Bern had just walked into the Guard-room and quietly given himself up.

Bern told the Commandant the following morning that he very much regretted having killed the Corporal, and that it was only the thought that he might be forced to shoot other men before he could get the youngster that had led him to surrender to the Guard. As he returned to the Guard-room with his escort after leaving the Orderly Room he passed the youngster on the road. "You rotten swine," he shouted hoarsely, "If I had my rifle now, I'd still fill your body full of lead, the same as I swore I would."

If a soldier committed a murder within sixty miles of a civil court he came under the jurisdiction of the civil authorities, but if the crime had been committed outside that radius, as was the case now, he was tried by the military authorities. In due course Bern was court-martialled, found guilty of wilful murder, and sentenced to death by hanging. As there was no prison at Kilana or Chakrata we wondered where they were going to hang him and who was to be the executioner. The military authorities were responsible for these arrangements, the same as for the trial. Behind the Guard-room was a very steep track leading down the khud-side. About a mile down this track was a small plateau on which some natives, under the supervision of the chosen executioner, started to erect a scaffold. The executioner, who was a Government hangman, was supposed to be a half-caste, but if he had even a splash of white blood in him it must have been hidden by the seat of his pants. This caused a lot of angry comment among the troops: everybody was saying what a damned disgrace it was for a white man to be hanged by a native. The murdered Corporal by this time had been forgotten and every man's sympathy now went out to Bern. Scores of men remarked that they would willingly pull the lever that would send him to his death rather than let a dirty black soor do it.

After sentence was passed, Bern turned religious. He declared that he forgave the youngster who was the cause of his trouble, and spent a lot of his time in reading the Bible and singing hymns. The parson visited him daily; also most of the wives of the married crocks were tender-hearted and brought him all manner of luxuries. He lived like a lord and gained over a stone in weight during the time he was waiting either for a reprieve or for his execution. The Guard-room was on the edge of the Square where football matches were played almost every afternoon; he was allowed to stand outside the Guard-room with an armed escort and watch the play. He never missed a match in which his own unit was playing, and he seemed to enjoy it as much as any other spectator.

No reprieve came through. On the morning of his execution the military authorities took careful precautions that no natives except the hangman and his assistant should witness the execution of a white man. Parties of troops with fixed bayonets were sent out on the khud-side early in the morning with instructions to allow no natives within a five-mile radius of the scaffold. They were to turn back anyone who attempted to pass them. In case of accidents more armed troops were posted on the trails and tracks up to one hundred yards distance from the open scaffold. These precautions were necessary, because every native in the

Bazaar and villages knew that a white sahib was going to be hanged that morning. A strong escort under the command of Major Lloyd of our own Battalion escorted Bern down to the scaffold, which he mounted, as firm as a rock. The escort had ringed the scaffold in. He wished them the best of luck, and five minutes later it was all over.

About a fortnight before his execution Bern had made an earnest appeal to the parson to be buried alongside the Corporal. The parson forwarded the request to the Bishop, who granted it. Bern was overjoyed: he said that he would now die a happy man. A new cemetery had been opened at Kilana and the first man laid to rest in it was the Corporal. Fortunately the ground had not yet been consecrated: if it had been, Bern's request would never have been granted, because a murderer who has been hanged is not allowed to be buried in consecrated ground. Shortly before the troops stationed on the hill left for the Plains, the Bishop arrived to consecrate the cemetery. The Bishop, who wore full dress at the ceremony, was afterwards referred to by some of the men as "Pretty Bo-Peep." This nickname was suggested by his bishop's crook and also by his having lost sheep on the brain, like the majority of parsons. It struck me that there was not much difference between this sanctification ceremony and some of the heathen ones I had witnessed since I had been in the country, except that it was not accompanied by so much awe and religious enthusiasm on the part of the congregation. After the Bishop had sprinkled the ground with holy water, the Don Juan, whom I mentioned before as having asked to do the extra sentry-go at Agra, and who was present at the ceremony, remarked to his pal: "Well, I don't know what you think about it, but I think this holy water business is a lot of tommy rot. If I had my choice I would far prefer having a thimbleful of a white girl's toilet-water sprinkled on my grave than have the whole River Jordan running in flood over it."

There were no murderers in the Royal Welch in my time, unless I count a certain married sergeant of thirteen or fourteen years' service who, one day, at the close of the following summer, went out for a gharri-ride and brought it to a sudden stop by shooting the gharri-driver dead with a revolver. The affair caused a mild sensation. The Sergeant always carried this revolver about with him, but had never threatened anyone with it or shown any murderous inclinations. He was a reserved man and respected by all ranks. Nor could anyone come forward, when the case was being inquired into, and give any evidence that the heat had been causing him to behave queerly of late. But the Commanding Officer allowed him the benefit of the doubt and had him confined in a small padded cell in hospital, where he was kept under observation. The gharri-wallah's widow came to the hospital to get justice done her. They explained to her there that the man who had shot her husband was a madman and that nothing could be done, except perhaps to give her compensation. They gave her ten rupees, which was a matter of thirteen shillings, and it is said that she went away well satisfied and smiling. We never discovered whether this compensation came out of the Sergeant's pocket or out of the Canteen funds. In any case, the excitement had completely died down a day or two later when the Sergeant officially recovered his reason, left the padded cell for the convalescent ward, and was soon afterwards invalided home to England with his wife and children. The truth never came out. It may be that the gharri-wallah tried to blackmail the Sergeant by threatening to tell his wife of some gharri-rides he had taken him on in the past; but that is only an unfounded theory. The general opinion was that the Sergeant's nerves were in a bad state after the heat of the summer and that the gharri-wallah must have demanded more money for the ride than what it was worth and given the Sergeant cheek when he refused; and that, with a revolver in his belt, the temptation to make a stern example of this cheeky gharri-wallah to warn all cheeky gharri-wallahs of the future proved too much for the Sergeant – he drew the revolver and ended the argument. So it was reckoned a good deed: for, as I have said, the Sergeant was not known as a bully or an unjust man, but was respected by all; and the gharri-drivers of Agra were certainly the limit.

The Battalion was, as a matter of fact, remarkably clear of crime, in the civilian sense of the word. This was not the case at Plymouth, where many a man's new shirt or a brand-new pair of boots were carried off from his bunk by unknown hands, taken into the town and sold cheap to civilians. Blankets were also stolen from the store-rooms in the evenings and disposed of in the same way. I do not remember any thief being caught, but if one had been, his room-mates would have put him through the mill to some order. This sort of thing did not happen in India. There we did not have so many young soldiers about with undisciplined habits left over from civil life, and living was cheaper there, which removed much of the temptation to steal, and stolen goods were not so easy to dispose of; but what accounted most for those goings-on at Plymouth was that ours was a composite battalion. For one thing is sure: mix men of different regiments together and crime always goes up. In the Army it is always considered more excusable to "win" or "borrow" things belonging to men of other companies, than things belonging to men of one's own company, and if a man gets a chance to pinch something belonging to another battalion, with whom his own is on bad terms, it is considered a good joke if he gets away with it. In war-time, of course, theft increases a thousand-fold: when one thinks nothing of taking a man's life one gets careless about relieving him of less important possessions. But no man worth anything steals from a comrade either in war or peace; which is the rule in my native coal-mines as well as in the barrack-room. I only recall one case of theft in India: it concerned a man in my company at Meerut shortly after I joined the Battalion. He was proved to have stolen a comrade's cap while the Battalion was at Hong-Kong. He was sent on from Company Orderly-Room to Battalion Orderly-Room, where the Commanding Officer refused to give him a summary punishment but referred his case to a District Court-Martial, which gave him a decent prison sentence, and deprived him of his North China medal. He was considered to have brought dishonour on the company. During certain stages of the War, however, caps, puttees and almost everything else but a man's rifle and gas-mask were more or less common property. As for bigger things, our transport horses had been practically all stolen from other battalions or corps, or from the French.

I had lost three stone in weight during my illness, but got it all back in the course of my stay at Kilana. When I rejoined the Battalion in November I felt in magnificent condition.

In the Autumn and Winter months the Battalion teetotallers, who were fond of dancing, occasionally organized dances in the gymnasium. The dance-band consisted of two signalers who were expert performers on the mandoline and banjo, and each charged ten rupees a night for playing. All the married crocks were invited to these dances and also a number of young half-caste ladies who lived in the civil cantonment and were generally accompanied by their mothers, who kept a watchful eye on them. Most of these girls had been decently educated. Some were as brown as natives, some were a half-and-half colour, but a few were as white as any European. They much preferred a pure-bred white husband to one of their own breed and their parents were generally able to give them a marriage dowry of between five hundred and a thousand rupees. They looked for husbands among men who were staying on in India – such as those who intended to become railway employees. One railway company invited men of good character to go through a six months' course of railway training. The only men who were eligible for this course were those who had six or twelve months to serve before they became time-expired. If they went through this course they were away from the Battalion for six months but drew their ration-money and pay; the railway company found them lodgings and also paid them a rupee a day for the duration of the course. At the end of it they had to pass an examination, but even if they passed it they could please themselves whether they took a job on the railway or went back to the Battalion. If they accepted a job, they were transferred to the Army Reserve, in spite of having perhaps six months still to serve before they became time-expired. They generally began work as passenger guards, and rose from that to something higher. The four men from the Royal Welch whom I remember going through this course all accepted jobs and all married half-caste girls whom they had picked up meanwhile.

But there were drawbacks to these marriages. Just before the Battalion left India for Burmah I met one of these four men. He had married a half-caste girl who was whiter than he was, and nobody in England would have taken her for an half-caste, or "half-chat" as the troops in my time contemptuously called them. It was not a term I used myself, any more than I would speak of "barrack-rats," for the white children. They called themselves Eurasians, a word formed from combining "European" with "Asiatic," but both Europeans and Asiatics despised them. It was a case of the parents eating sour grapes and the children's teeth being set on edge – even unto the second and third generation. This man had been married two years and told me that he was now the father of a baby boy, twelve months old, who was a throw-back to his wife's native ancestors. He said gloomily that the child seemed to be getting blacker every day. It told him it was tough luck. "Tough luck!" he echoed. "You have no notion how tough it is. I'll tell you candidly that I would not be ashamed to take my wife to Blighty to see my family, but I would be thoroughly ashamed to take my child. I have thought it all out. Suppose I took my wife and kid there, not one of my relatives would believe that she was a half-caste; but when they saw the kid they would not be backward in telling me that I was a damned fool to believe that I was his father. I don't regret marrying my wife, but I do regret being the father of my kid. If he lives I am doomed to spend the rest of my life in India. She often blames me for not taking more interest in him. I can't hurt her by telling her why I don't. At any rate I am determined not to be the father

177

of any more: I have used contraceptives ever since. I shall do my best for the kid and in course of time he'll grow up, I suppose, and get married. But by that time I shall be taken for a half-caste myself." He was living at a place called Tundla, which was a big railway centre. There was quite a colony of half-castes there.

Lord Kitchener had introduced what was called the Kitchener's test, which every infantry battalion in India had to carry out once a year. The first time we did it was when we were at Agra. It was a severe experience, but a true test of the marching and shooting powers of an infantry battalion. He offered a cup which would go to the battalion which was awarded the most points. The only men who did not take part in these tests were the hospital cases. About nine hundred of us paraded in fighting order and each man was issued with so many rounds of ammunition and so many squares of lead, that fitted into the pouches. The squares of lead represented more rounds of ammunition, bringing the total weight up to the equivalent of one hundred and fifty rounds. Each man was also issued with iron rations, which was all the food he got that day, consisting of four army biscuits, a tin of compressed soup and three blocks of chocolate. The test consisted of a fifteen-mile march followed by an advance for a mile in skirmishing order to targets dotted here and there in the jungle. As soon as the targets were spotted we had to find our own range and open fire. The advance was made by short, sharp rushes, and at a given signal all firing ceased, whether we had expended our ammunition or not. After an imaginary charge at the targets there was a retirement back to the spot where the water carts were, a mile away. We were then given one hour's rest, during which we lit fires and boiled up water for our soup and chocolate. By the time we had got down to that meal, with the four biscuits, and smoked a fag, the hour was up, and we marched back fifteen miles to the starting point, which was three miles from Barracks. Before the Battalion arrived back at Barracks they had done their thirty-eight miles since morning. Only five men fell out, all on the return journey, and all staunch teetotallers. I honestly believe that if they had been fonder of beer than they were of tea they would have stuck it. Their names were posted up in a prominent place in the Library, so that everyone should know who the weaklings were who had lost points for the Battalion. They were also punished by having to do an extra hour's drill every afternoon for a month. It was an unwritten law in the Battalion that no man should fall out on any march unless he had an attack of fever or ague; any man who fell out with blistered feet or through physical exhaustion was looked upon with contempt and deemed unworthy to belong to the Royal Welch. We did not win the cup but were runners-up to the winners, the Queen's (Royal West Surrey's) who beat us by a narrow margin. It was the Queen's who relieved us at Agra at the latter end of 1907.

To be a member of the Army Temperance Association a man had to sign the pledge and contribute four annas a month to the funds, which gave him the use of the A.T.A. room, where there were a number of newspapers and magazines not to be found in the Library. Members who kept their pledge were entitled to A.T.A. medals. One was awarded after six months, and I believe that they continued to be dealt out at yearly intervals, but a man would have to be back in civil life before he could appear in public wearing these certificates of unnatural behaviour. Every six months the members elected a new barman and secretary, who were each paid £ 2 a month out of the funds. These were coveted posts, because if an additional £ 20 or £ 30 a month could not be made between the two of them, and a handsome profit still shown, this meant either perfect incompetence or, what was less likely, perfect honesty. A percentage of the profit shown was sent to the Headquarters of the A.T.A., the remainder was spent on card-tournaments, dances and concerts. I knew about the money side of the business from one barman who, during the time he was on the job, often held cheerful card-parties in his bunk after eleven o'clock at night: they were made cheerful by the whiskey and soda which he provided for the party. He was very fond of a whiskey and soda, though the staunchest of teetotallers in other respects. Tea, which we called "char," mineral waters, cake, and bread and butter were sold at the A.T.A. The main profit was made out of the char, which was sold at two annas a pint. Genuine bun-punchers or charwallahs, as they were called, would drink char all day: summer and winter they drank it. The Prayer-wallah often said that if the bodies of all the A.T.A. men who had died in India from causes known or unknown were to be assembled in one grave, the finest tea-garden in the world would soon sprout up from between their bones.

HEAT

Towards the end of March 1906, the half-time party left for the Hills, but it was decided this year to keep the junior Signalling-Sergeant and six signalers on the Plains. As I had been away at the Hills for practically the whole of the previous summer I could not complain when I was told that I had been chosen to stay behind. The Prayer-wallah was one of those who went. For the first half of the summer it was so hot that no parades were done between half-past seven in the morning and sunset, though there was a short parade before breakfast and an occasional drill parade after sunset, during which we signallers did a little lamp-work. Throughout the months of May and June the day-temperature in cantonments kept between 116 and 120 in the shade, and 125 was frequently recorded at the Fort. At midnight the thermometer sometimes stood at 109. Day and night we were in a constant bath of perspiration. The two swimming-pools were not large enough to hold many men at a time and were always crowded. But then some men got diseases of the ear, which led to the perforation of the drums, through spending too long a time under water. When the cause became known, the baths were not quite so crowded as before. During the day most of us knocked about the bungalows with only a short pair of underpants to cover our nakedness; some only wore a towel tied around their loins. A great part of the time was spent in reading or playing cards, but most of the men, no matter what they happened to be doing, found time to curse the punkah-wallahs for not pulling the punkahs strongly enough or the tatty-wallahs for not throwing sufficient water on the tatties. The sun beat fiercely down and it was very hurtful for the eyes to stare for any length of time at the ground outside the bungalows. Occasionally a hot wind swept across the cantonments, which reminded me of the blast-furnaces of my native town, and which set any man who happened to be out on the verandah at the time gasping and choking. Many of the men lay out on the verandah all night as naked as they were born. Some used to go to the wash-house and sit in large zinc hip-baths full of water; they said that they slept longer without waking that way than stretched out on their beds under the punkahs. I never tried the hip-bath method myself, and also found it cooler lying on my bed under the swaying punkah than on the punkah-less verandah. "Cooler" is hardly the word I want: men lay on their beds twisting, turning and cursing for hours in their endeavours to get to sleep in the unbearable heat. If the punkahs ceased to sway for only a second somebody would shout: "Cinch, you black bastard, or I'll come out and kick hell out of you." I have already described how the two punkah-boys sat back to back on the floor of the small landing that divided the Signallers from the Police, pulling at the ropes for their respective rooms. If one of them thought that all the men in his room were asleep he might take a little doze himself. But there was always someone awake who would immediately shout at him to cinch, and if this warning, or the nudge of his companion, did not set him going again at once someone was sure to rush out on the landing and give him a running kick in the ribs. I never saw a punkah-wallah eat anything on his eight-hours' shift. His only refreshment was an occasional drink of water out of an earthenware vessel, called a chatty, which he brought with him and which held about three pints.

We were all covered from head to foot with prickly heat rash, which itches intolerably, and quite a number of men were troubled with boils and water-blisters. Water-blisters, caused by the heat, would sometimes rise on a man's face and when they burst his skin would be in a nice condition for a few days. I also remember thirty men being in hospital at the same time with water-blisters on their privates. By the end of the first half of the summer, even without the help of water-blisters, the whole Battalion had faces like whitewashed walls. We were no longer allowed at the Canteen at midday: it was only open during the evening. But at midday the section-corporal took the names of men who required beer and then escorted natives to the Canteen, who carried the beer back to the bungalows in chatties. Each man was supposed to have only a pint, but there were sixteen men in a section and if fourteen of them did not ask for beer, either because of being stony-broke or because of being bun-punchers, the remaining two could share the whole sixteen pints between them – supposing, that is, that they had the necessary money. The Canteen-Sergeant and barman did their best to keep the beer cool, but it always reached the bungalows lukewarm. Ice could be bought at the A.T.A., where there was a large ice-chest, but genuine beer-wallahs would have drunk their beer scalding hot rather than dilute it with ice.

Most men bought small native chatties for drinking purposes and kept them full of water beside their beds. These were wonderful vessels for keeping water cool but their drawback was that they were also wonderful breeding-places for mosquitoes. If a man had not drunk all his water during the night he would be sure to find mosquitoes in his chatty the next morning. Mosquitoes lay their eggs in water and the dirtier the water is the better it suits their purpose. One of the Black Pots told me that it was the bite of an infected female mosquito that caused malaria; the bite of a male, he said, was harmless. He may have been right; but just as the bite of a flea will cause much discomfort to the majority of human beings but will not trouble others, so with mosquitoes, male or female. After a man had been bitten he felt the bite like the sting of a small wasp, and it nearly always left a swelling. The finest black eye I ever had in my life was caused by a mosquito bite. I knew two men whom a mosquito would never bite; they claimed that they were mosquito-proof, yet both of them had been in hospital with malaria. The Prayer-wallah said that it must have been caused by mosquitoes spewing and making

183

messes all over them in their disgust at being unable to get a bite in. When I asked the Black Pot for his professional opinion on these two men, he said that they must have been bitten by an infected mosquito without being aware of it; otherwise they would never have had malarial fever. I told him that they had never had a swelling. He answered that this was quite possible, for the flesh of some people would not swell after a bite. But, going by my own experience, I shall never believe that anyone could be bitten by a mosquito without being aware of it; some of the bites I got would have penetrated the hide of an elephant and caused it to trumpet with rage.

The coolest bungalow in the Barracks was the Library, and the dogs soon found this out: it was very funny to see them trotting over from the other bungalows early in the morning and lying down under the tatties there. Sometimes they fought for a favourite tatty, and once in possession they would lie there all day and not depart until after the sun had set. There was a fine collection of books in the Library and I spent the greater part of the day there reading under one of the tatties. I now preferred straight history. I read Plutarch's *Lives of Illustrious Men*, Lord Macaulay's *Life of Cromwell*, and *The Life of Napoleon Buonaparte* by M. Bourienne (that was the author's name, I think). I found them more interesting than any novel I had ever read. A number of men spent their whole day in the Billiard-room, where they would play skittle pool, a favourite game with the billiard sharks. The three tables in the room were constantly in use and refreshments could always be had in the A.T.A. room next door.

We lost a number of men from heat apoplexy, enteric fever and other complaints. The first symptom of heat-apoplexy is being unable to sweat, and if a man does not sweat in a temperature of 120 in the shade he soon notices it. There was the case of Private George of the Regimental Police, who was five-foot-eight in height and weighed twelve-stone-six. He had never suffered from fever or ague, and I had even heard him say that he could not remember ever having had a day's illness in his life. One morning, just before dinner, I went into the Police room and found George lying on his bed, looking a bit queer. I asked him if there was anything wrong with him, and he told me that he felt burning hot all over and had not sweated a drop for about an hour. He had drunk a quart of scalding hot tea in an attempt to restart his sweating system, but without success. I did my best to persuade him to slip across to the hospital, but he said irritably that he was not going to go before a doctor simply because he felt hot and couldn't sweat. His room-mates had also been urging him to go to the hospital, and one of them now told him bluntly that if he did not make the move inside of half an hour he would be a dead cock before sunset. At about two o'clock that afternoon George lost consciousness and was carried to hospital, where he was found to have a temperature of 110. They put him on the slab, but although the Doctor and the Black Pots worked very hard, rubbing him down with blocks of ice in an attempt to reduce his temperature, it was no use: he died in less than an hour. A Black Pot told me that his life would have been saved if he had been brought across an hour or so sooner, and said that many a man had died in the same way through hanging on too long after feeling the first symptoms.

The Battalion lost between fifty and sixty men this summer, more than half of whom were H Company men, who all died in the same month in the latter half of the summer. The Company began to take notice, after a few funerals, that as soon as they arrived at the graveside a solitary crow would alight on the cross of a tombstone a few yards from them and stay there watching the proceedings. When the coffin was lowered into the grave the crow would start to caw, maliciously as it seemed, and even the three volleys that were fired over the grave did not frighten it away. Then, as the company left the cemetery, it would fly behind them, still cawing, until the last man had passed through the gates; after which it would fly silently away. If they did not happen to bury a man one day they buried two men the next, to make up for it; and the crow – they all swore that it was always

185

the same crow – never failed to present itself at the graveside. This so got on the nerves of the Company that they believed they would continue to lose men so long as the crow appeared. And the queer part of the story is that they were right. At the funeral of the thirtieth man the crow did not put in an appearance, and this was the last man the company lost for the rest of the summer.

The vegetable-wallahs and others sold their stuff at the Ration Stand, which was under some large shady trees in the centre of the Barracks. Fruit of all sorts was cheap, but the favourite with most of us was the mango. Mangoes are about the size of oranges, with a thick yellow-green skin which has to be peeled off and a large stone in the centre. They are delicious to eat but if they are not quite ripe will soon double a man up with colic. Every morning and late afternoon a native from the dairy would come around the bungalows with a pony and cart. He had an ice-chest in the cart which contained pats of butter which he sold at the usual price, but unless a pat was eaten at once it soon ran off the plate. Meat was issued at the Battalion Ration Stand a few hours after the cattle were killed, and cooked immediately. I ate very nearly as much meat in the summer as I did in the winter. At the coffee-bar, where one could drink beer all evening until stop-tap, some excellent suppers could be had. A large plate of cold meat with salad and a piece of bread only cost one anna; a cooked cow's tongue only cost two annas; and a large cup of oxtail soup one anna. A curious effect of the heat was that one could drink beer for hours without it having any effect on the bladder. One man I knew used to swear that he only eased his once a month, and that he did this as a matter of routine, after signing his monthly accounts, in order not to lose the facility.

There was a large theatre in the Barracks, and the travelling variety companies who hired it often gave us real good turns, well worth the eight annas which was the lowest charge for admission. Shortly after the half-time party arrived from the Hills a George Edwards touring company took the theatre for a week and played a different musical comedy every night. The lowest price of admission to their shows, which began at ten o'clock at night, was two rupees; but if a man did not have the money he could get a ticket from his Colour-Sergeant to attend any show, and the two rupees would be deducted from his pay at the end of the month. The theatre was packed every night. I attended every show and their performances of "The Earl and the Girl," "H.M.S. Pinafore," "The Cingalee," and "Sergeant Brue," as I remember them now, would have knocked spots off any of the modern musical comedies. What appeared to be a severe criticism of these performances was made by one of our corporals, a North China veteran who was soon to be time-expired. He was a jovial chap usually, but had not been himself for some little time. He walked out in the middle of a performance and was not seen again that night. The next morning he was absent from early parade and someone suggesting getting grappling-irons to drag the well outside his bungalow. He had been seen loitering about the well-head two nights before. Sure enough, they found his body there. He had apparently dived into the well head-first and struck his head on the brick-work, which stunned him before he entered the water; for the doctor could find none of the signs of a struggle for breath that one expects in the corpses of drowned men. It was not known why he had committed suicide – it was clearly suicide, not accident – but everyone exonerated the musical comedy company of all blame. The explanation most commonly given in the Battalion was that he had gone on the tact recently, saving up for civil life, and this change of habit had unsettled his health of body and mind. The beer-wallahs were strongest in expressing this view. They said that if he had kept to his drop of purge he would never for a moment have entertained the thought of suicide – his mind would have been too busy planning and scheming where the next midday or evening issue of purge was coming from.

We were occasionally troubled by a sandstorm. Towards the end of June a very severe one, which lasted for many hours, swept over the Barracks and blew down most of the tatties in the Library.

All bungalow doors were closed, which gave us a suffocating night, but in spite of this there was a layer of sand on the floor of every bungalow the following morning. The atmosphere after one of these storms was a lot clearer than after rain. The half-time party on their way from the Hills had to do a little cholera-dodging, a word which I shall explain soon. They were only a day's march from Dehra Dun when six fatal cases occurred, but luckily they had no more after leaving the camp where it had broken out. Because of this they rejoined the Battalion a week later than what they should have done. I was glad to see that the Prayer-wallah had not been one of the victims. He was as brown as a berry and commented on my washed-out appearance. Cholera was more feared by the troops than any other disease, and men who went down with it seldom lived longer than a few hours, dying in terrible agony from cramp in the calves, thighs and stomach. The general belief was that it was caused by a small invisible cloud that passed over at a height of two or three feet from the ground. Old-Soldier Carr, who had done a bit of cholera-dodging with the First Battalion, was firmly convinced that this was the case and explained that the cloud was packed with the germs of the disease. Whenever cholera was mentioned in his presence, he would tell the story of a man he knew in the First Battalion who had arrived back from the Canteen so drunk that he had fallen down by the side of his bed and gone to sleep on the floor. During the night a cholera-cloud passed through the bungalow, and in the morning he discovered that he was the only man alive in the bungalow; the remainder were dead in their beds. The cloud had missed him because he was below the level of its transit. Every unit serving in India had the same story: this miracle had once happened to one of their own men at some time or other. I would bet twenty rupees to a pie-piece that the same yarn is still spun today in every unit in India.

Being of an enquiring nature, I asked both the Bacon-wallah and the Abyssinian Warrior if there was any truth in this yarn. They both believed that it came over in a cloud but assured me, also, that no man would be safe from it even if he was sleeping in a trench twelve feet below the ground-unless, of course, he had taken a skinful of rum. Beer, they said, was not strong enough to counteract the germs. The Abyssinian Warrior recalled a personal experience during an outbreak of cholera in his battalion. He had woken up in the morning in a tent which held sixteen men and found only one other man alive, who happened to be his boozing chum. The two of them had drunk enough of rum, he said, the night before, for the cholera germs to die of stupefaction the moment they approached within a foot of their faces. The Bacon-wallah had not had this experience himself but could recall many similar cases in his own battalion. He said that the worst outbreak of cholera he could remember was about nine or ten years before the Mutiny. In forty-eight hours they lost over two hundred men and four officers, and then about thirty men more while cholera-dodging before it was finally stamped out. This was the more credible yarn of the two, for in Agra Cemetery there is a monument erected in memory of one hundred and forty-six men of a battalion of the Yorkshire and Lancashire Regiment, who were stationed at Agra in the middle nineties. They all died in less than forty-eight hours after the first appearance of cholera. Cholera-dodging was as follows: when an epidemic of cholera broke out in a battalion the whole battalion would immediately pack up and being marching against the wind. They would pitch camp after a full day's march. If a case or two occurred here, they would continue their march the next day, and keep on from camp to camp until the epidemic had been stamped out. They would not return to their original station until it had been well fumigated. A daily tot of rum was, I believe, issued to the men when cholera-dodging. It had been done in the time of the Bacon-wallah and also in the First Battalion when Carr was serving with it. Carr said that the Canteen had been kept open all day during the outbreak. With the exception of those half-dozen cases in the halftime party, the Battalion never had to contend with an outbreak of cholera all the time I was with them, and this good luck held for the rest of their overseas tour after I had become time-expired.

I have mentioned Old-Soldier Carr's name several times and I must now close my account of him. Let me declare that he was the smartest soldier I ever saw, both in action and in appearance. He stood six feet high in his socks and was beautifully proportioned into the bargain – which is rare enough in tall men. He left the Battalion for home shortly after we proceeded from Agra to Burmah, but he was not half the man then that he had been when I first knew him: for he had lately taken to drinking a native fire-water which we called Billy Stink. One could get it cheap in the bazaars, and it was a sort of wood-alcohol, I believe, though I never cared to sample it myself. Its effect on most drinkers was terrible: I knew two men who had to be put in strait-jackets in a padded ward through indulging in this stuff. Carr drank it without so much as winking, but it got him down at last, and, within six months of his leaving us, news came of his death. India killed many a man at long range. But better a thousand times to die of the after-effect of drinking Billy Stink than of the after-effects of going with a sand-rat, which was the fate of some poor devils.

About the middle of July there came a break in the weather – the day-temperature sank to between 108 and 110 in the shade, which allowed us to do light parades up to midday. For months the natives in the bazaars and the villages and the city had been making the nights hideous with singing, praying, the beating of drums and the firing off of maroons, in vain appeals to their gods to send them a shower of rain. About the middle of August the gods relented and rain began to fall in bucketfuls. We welcomed it as much as the natives did, and most of us Signallers got up on the roof of the double-decker bungalow, which had a concrete verandah on each side, about six feet wide, with a parapet running along it. We let the rain beat down on our naked bodies and found it wonderfully soothing for the prickly heat; quite a number of men from the neighbouring bungalows were standing outside in the open, doing the same thing. As at Meerut, huge water-snakes emerged from cracks in the ground, and disappeared as soon as the rain ceased. And yet the nine hours' rain had made little impression on the soil, which was still caked hard. The good that the rain did was balanced pretty closely by the harm. The stench that rose up afterwards was appalling and in less than a fortnight over a hundred men had been admitted to hospital with fever, in most cases malaria. At least another hundred were hanging on in their bungalows with lesser attacks of fever and ague, in the hope that they would be all right again in four or five days' time. As there were no parades in the afternoon this could be managed. It was not an uncommon sight to see men spending their afternoons in bed with a pile of blankets and top-coats heaped over them. They took quinine tablets and soda-water mixed with milk as a means of making themselves sweat. Their beds would be shaking like boats in a rough sea for a few hours, but, once they began to sweat, the fever and ague left them. They would then be all right until the following day, but at exactly the same time as the fit had seized them on the day before they would have a recurrence of it. If it did not leave them by the fourth or fifth day they would go sick. At this time, for about a month, every man in the Battalion had to parade twice a week and swallow so many grains of quinine. I thought this was a good idea, though there were some who hated quinine like poison; but they could not get out of taking it because the Doctor saw that every man swallowed his allowance down before he was dismissed off the parade.

The evening after the rain-storm it was found impossible to play billiards. Winged insects of all shapes and sizes from a hundred miles around, it seemed, had invaded the Billiard room, making for what they mistook for grass on the billiard tables. They were so thick that the balls would not run more than a few inches through them. The native markers brushed off and killed thousands of them, but thousands more instantly took their places. They had all disappeared by the following morning.

With the exception of minor discomforts I had enjoyed very

fair health throughout the summer, but early in November, after hanging on for four days, I was admitted to hospital with a recurrent attack of malaria and was again put in the special ward. This attack was not so bad as my first one. My temperature went up to 105 for the first two days but afterwards never rose higher than 102 in the afternoons and evenings. In less than three weeks I was all right again and eating like a horse. After this experience of hospital I came to the conclusion that the old Colonel was indeed as stone-balmy as everyone said. For the time being he had forgotten about Delhi Fort and flies and now had influenza and tonsillitis on the brain. He came around the ward one morning and three of us who were marked up as "out of bed" stood rigidly at attention at the foot of our cots. By the time he had arrived at mine he had told the ward doctor that the other two had not had a recurrent attack of malaria at all: they had simply had a bad attack of influenza. He eyed me up and down for a few moments and told me to open my mouth. He then seized hold of my tablespoon, which was on the locker by the side of the bed, and with a triumphant look on his face inserted it in my mouth, pressing down my tongue.

"Say 'Ah,'" he ordered.

Now, a regulation Army tablespoon is a lot larger than any ordinary tablespoon and with the end of it wedged against my tonsils I found it impossible to say even "Ah."

He withdrew the spoon and after another careful look at my mouth turned to the ward-doctor. "I knew it, I knew it, I knew it. This man has had a bad attack of tonsillitis and still shows traces of it. It's marvellous how well he has got on, considering that he has been given no treatment for it."

I don't know what the ward-doctor's thoughts were, but he did not reply. Perhaps he was thinking that it was only a question of time before he himself would rise to the rank of Colonel and become just as balmy himself.

The three of us were transferred to the convalescent ward that evening and when the doctor of that ward saw us the next morning

192

he did not know what to make of us. He picked up our boards and saw one disease crossed out and another written underneath it. Naturally he wanted to know what had really been the matter with us. When the other two told their story he did a grin; when I told mine he did a larger grin, and after examining my tonsils he did a very large grin indeed. But he made no comment. I then told him that I felt fit enough to return to duty and asked him to mark me out of hospital. He slyly replied that he could not do that, because the Colonel had found that I had still traces of tonsillitis. He added that another fortnight in hospital would do me the world of good. There was one thing that the old Colonel always believed in, whatever his latest insanity might be, and that was that his patients should be well fed. My own diet was a very liberal one and I was now allowed a pint of beer for my dinner. The Prayer-wallah had also been bringing me in two pints of beer every night: it was very easy to smuggle beer into this hospital. So I had no reason to complain.

When a man was marked out of hospital by his ward-doctor, he had to appear for his final discharge in front of the old Colonel, who generally gave him a lecture on how to avoid the complaint of which he had been just cured. My fortnight in the convalescent ward had lengthened to nearly three weeks before the ward-doctor marked me out. I then appeared in front of the Colonel, who kindly inquired if I experienced any trouble with my tonsils. I replied that, so far as I remembered, I had never had any trouble with them in the whole of my life. The Colonel half rose up out of his chair and eyed me so fiercely for a few moments that I fully expected him to summon two hefty orderlies to hold me down by force while he extracted my tonsils there and then. My reply had so enraged him that for the time being he seemed incapable of speech. Then he slowly sank back in his chair and began to mutter that in the whole course of his career as a doctor he had never met so ungrateful, so insolent and so ignorant a young man as I was. After having been cured of my tonsillitis by all the best means known to modern medical science, and having had every care and attention lavished on me, I now had the damned impudence to inform him that I had never suffered from this complaint. His voice gradually rose up to a bellow and he kept on at me for a few minutes without a pause; but I had the sense not to answer him back. If I had done so, he might have discharged me from hospital with the crime of "insolence to the Principal Medical Officer of the Station"; which would have landed me in the Battalion Orderly Room and the least punishment I could have then expected would have been seven days in the cells. Still, to give him his due, he had never yet been known to crime any of the men whom he had considered to have been insolent to him, the though he had threatened one or two of them with life sentences. After calming down he gave me a lecture on the origin, diagnosis and treatment of tonsillitis, and told me in the conclusion that if I was ever troubled with my tonsils again the best thing I could do was to have them extracted. I thanked him and he then dismissed me. This was the last time I was in his hospital as a patient.

The ward-doctors whose care I had been under were capable and humane men. One of them, Captain Scott-Worthington, had been more successful in his operations for abscess on the liver than what the others had been. Men for who went sick with this complaint considered that they had quite a sporting chance of living if he operated on them.

YANK

I have met some queer men in my time but the queerest of them all was a man who had enlisted a few days before me and went by the nickname of Yank. We were at Wrexham and Plymouth together and then he was sent to South Africa with the cease-fire draft, joining the Second Battalion in India at the same time as myself. He said he had been born in England but had spent most of his life in America. The very day he landed at Liverpool on his return to his native country he had joined the Army; and that was about all that I or anybody else ever knew about his past. At home he indulged moderately in a smoke, a drink and a woman, but since his arrival in India he had denied himself all these luxuries and was now subsisting on the most frugal fare that would support life. He was a student of the occult sciences, which I believe was the motive that had led him to join the Army: he wanted the chance of spending a few years in India, which is the best place to study them, at no cost to himself. He had risen to the rank of full corporal on the Signallers and now occupied one of the bunks at the end of the verandah outside our room. His speciality was hypnotism. There were half a dozen men whom he could hypnotize at any time he pleased, and one of them, a drummer named Lewis, was completely under his control. The power he had over this man was unbelievable.

The Prayer-wallah and I were very friendly with Yank and occasionally visited his bunk after stop-tap to watch him performing. We twice saw him put Lewis into a hypnotic trance. As soon as he went under, Yank laid his rigid body across two chairs set at some distance apart. He then seemed to concentrate on the body, at the same time making gentle passes with his hands over it: gradually it began to rise in the air until it was suspended, still rigid, about two feet above the chairs. After encircling it with his hands about half a dozen times he made some more gentle passes and caused it gradually to sink again until it rested on the chairs once more. He then lifted Lewis off the chairs and placed him on the bed and speaking in a very sharp voice ordered him to wake up. Lewis was under the impression that he had fallen asleep on the bed. On these occasions he would never believe what had happened to him, because he never remembered anything about it. The first time that the Prayer-wallah and I saw Yank levitate Lewis, as he called it, we seized hold of two signalling flags and as soon as Lewis had come to himself we began to beat the air with them over the chairs. But the sticks came into contact with nothing. Yank smiling said that we had not witnessed any conjuring trick, only a short demonstration of the power of mind over matter. We were not satisfied with this explanation, so the next time we saw it done we swished the sticks all around the bunk before Lewis was placed across the chairs; but we found no wires or anything else from which he could be suspended.

It was very rarely that Yank would give a turn at a concert, but when he did appear at one, which was held at a Roman Catholic Soldiers' Home called the Paddies' Club, he not only hypnotized some of our own men but four men of the Royal Field Artillery as well. He made them do all manner of foolish things and concluded his performance by suspending Lewis in the air above the chairs. This time he had borrowed a sword, with which he cut the air above him and beneath him. Most of the men who witnessed this said that it was a damned clever trick and nothing else. But if a native fakir had performed it, no doubt they would have boasted afterwards of having seen real Eastern magic. Yank once told me very seriously that he was not studying the occult sciences for the purpose of making money. Any man who did that, he said, would not get very far with the business. He had won scholarships in Hindoostani and Persian and by the time we left India for

196

Burmah never used an English word when speaking to a native. He seldom mishandled one of them, but the majority were very much afraid of him.

Just after I came out of hospital our young cleaning-boy was taken ill and his father was given permission to bring another boy to assist him until his son got better. The boy he brought was an evil-looking man of about thirty-five years of age; he had been given a permit by the Quartermaster on the recommendation of the elder boy. The new boy, who was called Abdullah, had only been working a few days when he gave Yank a lot of cheek outside the bunk. Yank, instead of speaking to him in his own language, so that he would have trembled and shaken as many a native had done who had dared to answer Yank back, lost his temper and gave the man such a thrashing with a stick that he was sore for days after. I jokingly asked Yank why he had not hypnotized Abdullah before giving him the hiding. Yank replied, quite seriously, that there were certain persons whom it was impossible either to hypnotize or to hurt except with physical violence, and Abdullah was one of them. About a week later Yank decided to have a small grass-snake as a pet, but he did not want one longer than a foot or eighteen inches, he said. He asked the Live Hare-wallah to get him one but the Live Hare-wallah replied that they were not in his trade. Yank now asked the two cleaning-boys if they could get him the snake he wanted, and said that if either of them did he would pay him a rupee for it. Abdullah said that he knew a man who could capture any kind of snake and that if he saw him that evening he would mention the matter to him. The following morning he told Yank that he had seen the man, who would do his best to capture a small one for him. Two days later, which was a Saturday, Abdullah said that he had to meet the man that evening after sunset; the snake had been caught, but unless he was paid the rupee the man would not hand it over. Abdullah then asked if he could bring the snake that evening instead of next morning. Yank said that he could, and that, if he did, he would get two annas buckshee for his trouble.

In anticipation of the snake's arrival Yank had a small wooden box made at the Pioneers' shop, with a sliding glass side and holes punched in the top. He also bought some milk to give it a feed, as he thought it would be hungry after its capture. The two cleaning-boys went off at five o'clock and as there was no lamp work on Saturday nights our room was more or less deserted by six o'clock. At twenty past nine, when the Prayer-wallah and I returned from the Canteen, we looked in at Yank's bunk and found him crooning over a small and very prettily marked snake which was lying on his bed. It was about fifteen inches in length and had the brightest and wickedest-looking eyes that either of us had ever seen in a snake. Yank said that Abdullah, who had arrived with it in a cloth just after seven, had told him that according to the snake-catcher it was a very rare variety of grass-snake and dirt-cheap at a rupee. It seemed listless at first but after a feed of milk it grew lively and in less than no time Yank was handling it as if he had had it for years. He said he was so pleased with it that when Abdullah had left he had given him four annas instead of the two annas he had promised. He invited the both of us to play with his new pet, but we refused. The Prayer-wallah and I shared the same opinion: that grass-snakes as well as poisonous ones should be dead before they were handled.

The following morning Abdullah did not turn up for his work. The old cleaning-boy said that he had not seen him since they had parted in the bazaar the evening before, when Abdullah said that he was going to meet the snake-catcher. He could get through the cleaning work by himself for once and his own son would be well enough to work the following day, though Abdullah did not know this yet. That morning Yank showed his pet to the old cleaning-boy, the sweepers and the cooks, asking them if they knew the Hindoostani name for this species of grass-snake. Not one of them had seen one like it before, but the number-one cook said it looked to him like a barbary wallah; meaning a dangerous snake. Like the Prayer-wallah and myself, none of them would take hold of it, which made us think that there must be something not quite right about it after all. But Yank only laughed and said that he could not understand how some people were afraid to handle a grass-snake although they knew that it could not harm them.

For the last twelve months we Signallers had been working on Thursday mornings – Thursday was a general holiday – which exempted us from Sunday Church-parade. Just after the Battalion had moved off to church I persuaded Yank to bring his pet along to a man called Ogden, who worked in the Canteen, for his expert opinion. Ogden, who had fifteen years' service, had transferred to us from the Rifle Brigade the second year we were at Meerut. He was an authority on snakes and could reel off the names of the different species and varieties as easy as a child with its A.B.C. Yank put his pet in his handkerchief and off we strolled. We came across Ogden outside the Canteen, and Yank explained the reason for his visit. He took his pet out of his handkerchief and displayed it proudly.

As soon as Ogden saw it, his face changed colour and he took a few quick paces backward. Then he said: "For Heaven's sake, Corporal, don't ask me any questions, but do exactly as I tell you. Put that thing back in your handkerchief. Now lay it under your heel and crush it to death."

Ogden's frightened face put the wind up me, but Yank kept his nerve. Without a word of protest he did exactly as he was told. When he removed his heel we found that he had crushed the little snake's head to a pulp. After Ogden had examined the corpse he said: "Well, Corporal, I don't know how the hell you are still alive but I am very glad that you are. I have handled live cobras in my time but I would not have attempted to handle this pretty little thing, not while it was alive, for all the wealth in the world. Look, it's no thicker than my little finger but it's one of the deadliest creatures in the whole of India. If it had bitten you, you would have been dead in less than two minutes. It's rare, too. I first heard of the species about ten years ago from an old snake-charmer, but I never saw a specimen until today. I recognized the markings at once, from an illustrated book on snakes I have, that cost me a lot of money to buy." He went inside the Canteen and brought the book out to us. Then he found us the picture of the snake (the name of which I have forgotten) which corresponded in every detail with the one that had just been killed. The writing underneath said: "One of the rarest and deadliest of Oriental snakes."

Yank now said: "Well, Ogden, I had honestly thought it was a grass-snake, but I saw at once from the fear in your eyes that I was mistaken. I don't believe it would ever have bitten me, but what gets me is that I very nearly sent Dick and Prayer-wallah to an early grave when I asked them to handle it last night – to say nothing of those natives this morning."

Ogden now asked Yank how he came to possess the snake, and Yank told him the whole story, not forgetting the hiding he had given Abdullah. Ogden whistled under his breath and said that Abdullah evidently knew more about snakes than he had been given credit for. His opinion was that Abdullah had probably captured it himself, because it was not the sort of snake that one could buy from an ordinary snake-catcher, and that it must have been in a half-drugged state when it was brought to Yank, which was why it was so listless. Abdullah had evidently planned to bring it during the evening when he knew that most of the Signallers would be out; he also knew that the milk would revive it and he then fully expected it to bite Yank, who would die before his eyes – and this would have made his revenge sweeter. As it was dark at the time he could probably have vanished without a soul noticing him. He might have tried to get hold of the snake again, but this was doubtful: he would probably have left it to make its own escape.

Yank said: "Abdullah went off when the snake began to get lively."

"I can quite believe it," said Ogden. "He must have had the shock of his life when he saw the way you were handling it, and by that time he was in danger of being bitten himself. I hope he gets caught now; but the odds are against it."

Our Regimental police combed the bazaars, and the native police also took a hand in the matter, but Abdullah had completely vanished; nor could any native be discovered who was likely to have provided him with the snake. It was found that he was not a native of this part of the country and had only been living in the Regimental bazaar for about three weeks. He also had served several terms of imprisonment in different parts of the country for acts of violence, though our old cleaning-boy had known nothing of this when he recommended him to the Quartermaster.

So Abdullah got away with it, but our old cleaning-boy swore by his gods that he would do battle that evening with the man who had introduced Abdullah to him as a pukka cleaning-boy. He was in such a fierce mood all day that the natives working about the bungalow gave him a wide berth. The Prayer-wallah and I knew what the battle would be like, having been present at one or two similar ones before, so ten minutes after he had left for home we walked down to the Regimental Bazaar, which was built on one side of a very wide road. Soon a crowd of natives collected in a ring on the road, and our old cleaning-boy and his opponent took up their positions inside it, standing four paces apart, each with his arms folded on his chest. It was at this distance that they began and finished their battle. I had by now picked up a fair knowledge of the crab-bat, but I could not by any means understand all they said. The Prayer-wallah had to do a little quick interpreting, now and then. First the old cleaning-boy let loose his broadside and banged away until he temporarily ran short of ammunition. Then his opponent replied, with shot for shot. Then it was the old cleaning-boy's turn again. They called each other all manner of far-fetched names and used swear words and oaths that made the Prayer-wallah tremble with admiration. He memorized great strings of execration for his own future use.

As this fierce battle went on and on, the spectators were worked up to a terrific pitch of excitement and ecstatically applauded each foul spurt of angry abuse. They each went along the other's pedigree, generation by generation, making more and more loathsome discoveries, until our cleaning-boy was finally acclaimed the victor. He had gone back two thousand years in his rival's genealogical line and given convincing proof that a direct female ancestress had secretly cohabited for years during her widowhood with a diseased bull-frog, thus going one better than her mother, who had legitimately married and cohabited with a healthy pig. The loser slunk away from the ring, a beaten man. He had put up a great fight, according to Bazaar standards, but he had only been able to go back nineteen hundred years in the cleaning-boy's family history. The victor strutted off, as proudly as a man who has won the heavyweight boxing championship of the world. When he turned up for his work the next morning he was still flushed with victory.

As for Yank and his hypnotism, the last time he did a turn on the stage was at Kilana in 1907. Lewis was with him and a discussion afterwards arose among the officers of the different units stationed there, who were all messing together in the same building, as to whether Lewis was genuinely hypnotized or merely a clever confederate. The Commandant of Kilana gave them permission to clear up the point by inviting Yank down to the Mess, with Lewis, to give a show. When Yank arrived, a couple of senior officers took him aside and explained that the members of the Mess wished to assure themselves that it was genuine hypnotism, not trickery; and that therefore, after he had finished with Lewis, he had their permission to hypnotize, if he could, the junior subaltern in the Mess, who would be sitting in such and such a place and who did not know that he had been selected for the experiment. Yank agreed, and after giving his usual stuff, he asked the young officer if he would mind stepping forward for a minute. In less than no time he had hypnotized the officer and was making him perform the silliest tricks imaginable, of the sort that would have better suited the Canteen on pay-night than the

Officers' Mess at a scientific court of inquiry like this. The officers were greatly impressed, and also greatly amused, until it suddenly dawned on them that a grave breach of discipline was being committed – a mere signalling corporal was making a public laughing-stock of a commissioned officer. The same two who had called him aside now sternly ordered him to discontinue his experiment and return the young officer to his normal condition. Yank did so, but it was clear that the senior officers were cursing themselves as damned fools for their thoughtlessness: and within a few hours every soldier on the hill knew the story, together with the officer's name and regiment, and so did all the natives in the bazaar as well. Yank surprised the officers by refusing to accept any payment for the performance. He told me later that if it had not been suddenly cut short he would have been tempted to go a little farther: the two senior officers were themselves easy hypnotic subjects.

THE AMIR, AND GERALD

That winter the Amir of Afghanistan paid a visit to India. He was booked to spend a week at Agra and great preparations were made for his arrival. There was going to be a big display of troops for his benefit, and early in January 1907, white and native battalions were arriving daily, from all over the country. Camps sprang up like towns and a magnificent one was erected for the Amir himself, who besides his Ministers and personal attendants was also said to be bringing with him a small army of two thousand picked men. Much money was spent on this camp, which had a fine arched entrance lit up at night by hundreds of small electric lights. Inside the camp there were artificial trees and large tropical ferns, and a wall about three-foot-six in height encircled it. About a thousand yards away another camp was built, to accommodate Lord Kitchener and his staff. When the Amir arrived we found a guard over him of three officers and one hundred and seventy-six other ranks. As usual the tallest men in the Battalion were picked for this guard, and I was one of them. For three weeks we rehearsed daily for hours so that no mistakes should be made on a guard of this importance, and by the end of that time we were all so thoroughly fed up that our curses were loud enough to have been heard all the way to the Amir's Royal Palace at Kabul. We had two sentries posted outside the guard tents, one sentry on each side of the arched entrance, and a number more around the wall of the camp. The Amir also had his own guards inside the camp, so he was pretty safe.

When the Amir accepted the Government's invitation to visit India, the Viceroy and his council did not know whether to receive him as an ordinary Native Prince or as the ruler of a foreign country. King Edward solved the difficulty by sending a personal telegram, as Emperor of India, welcoming His Majesty the King of Afghanistan to his Empire. This was delivered to the Amir when he arrived at Peshawar. Since we were on guard over Royalty the only two people entitled to our services were the Amir, and the Viceroy, who represented King Edward. Lord Kitchener, though Commander-in-Chief of the British Forces in India, did not come into the picture, even when he visited the Amir's camp. The Amir, who seemed to be some forty years of age, was about five-foot-six in height and inclined to be a little portly. When he drove out in his carriage he was generally dressed in a black European suit, with a red fez on his head. His complexion was a good deal lighter than that of his fellow-countrymen: with his slightly hooked nose and black whiskers he looked like a well-to-do Jewish pawnbroker.

He seems to have possessed a grim sense of humour. The story going the rounds was that when his subjects learned that he was about to visit India some of them said that the Indian Government was not to be trusted, and that he would be assassinated as soon as he arrived at Peshawar. One man, who had the reputation of being a bit of a prophet, travelled around from village to village, preaching this yarn. The Amir had him arrested and kept in solitary confinement until he should decide what punishment to mete out to him. On the day before he started out on his journey he ordered the Royal Head-Tailor to stitch the prophet's lips tightly together with good strong thread. After this was done he had the prophet placed in a large wicker basket which was then hoisted into the fork of a high tree. He then left orders with certain officials, who were not accompanying him on his visit, that if they received authentic news of his assassination they were to release the prophet and reward him handsomely; but if they received no such news some of them would lose their heads unless the basket and its contents were still in the tree when he returned. The prophet died a speechless and horrible death. It was said that the Amir passed similarly far-fetched sentences of death on a good many criminals, yet he was one of the principal heads of the Mohammedan faith in the East and considered a wise and humane ruler.

In addition to ceremonial visits from Lord Minto, who was then Viceroy, and Lord Kitchener, the Amir was also greeted by a few Mohammedan Princes. One afternoon I was one of the sentries at the arched entrance, when an uninvited visitor arrived. He was a bearded old chap and came trottting across the open maidan mounted on a very clean racing-camel. He had a long curved sword at his side, and a bow slung across his shoulders, and a quiver of arrows at his back, and looked like Ishmael or Esau or some similar character out of one of the Five Books of Moses. He halted his camel at the gate, and speaking in a curious sort of Hindoostani asked us whether the Amir was in Camp. When we told him that he was, he seemed delighted. He said that he had been travelling for many days to pay his respects to the Amir. We had a difficult job to understand him, but he had a still more difficult job to understand us when we told him that we could not allow him to pass us without a permit. I did most of the talking and finally made him understand that he could not enter the gate, but only by bringing my rifle to the ready with the bayonet about a foot from the camel's chest. The Captain of the Guard, who was outside his tent, now came over to us and inquired what all the trouble was about. When we explained, he too told the bearded one in Hindoostani that nobody could see the Amir without a written permit; but gave him instructions where to apply for one. The old chap now became very excited and began to argue the point; so the Captain warned him that unless he cleared off at once he would be arrested. He finally went off, muttering in his beard, but when at some distance he turned his camel around challengingly and faced the camp. He unslung his bow and, quickly fitting an arrow to the string, shot high into the air in the direction of the Guard-tents. The arrow fell only a little short of the sentries there. He then wheeled about again and soon disappeared over the sky-line. After travelling hundreds of miles to no purpose for it was clear that there were personal reasons that prevented his applying for a permit – his howling rage was understandable. No shots were fired after him.

The Amir's picked troops were a better joke even than the Portuguese Expeditionary Force that took over trenches from us in the Bois Grenier sector during the late War. They were a dirty scuffy ragtime lot and never left camp for a short march without their own brass band. The big-drummer, who had a bobbie's job, would have made a cat laugh. Unlike any other big-drummer that I have ever seen or heard about, he made no attempt to carry his own instrument even when the band was not playing: the big drum was securely fastened on the back of the man in front of him. When the band struck up, the big-drummer banged the drum so hard with his two drum-sticks that it looked as though he had a personal grudge against the unfortunate man who was carrying it. At sunrise and sunset the band played in the camp and the hideous row they made should have been enough to make the most fervent Mohammedan change his faith. During a visit to the Taj Mahal the Amir noticed some small children, belonging to the Battalion, playing inside the grounds. One pretty little child, the daughter of the Regimental Schoolmaster, so interested him that he asked for her to be brought to him. He patted the child's cheeks and then clasped a necklace of pure gold nuggets around her neck. The Schoolmaster had many good offers for the necklace that afternoon and later, but refused to sell it. One day the Amir took the salute at a march past of about forty thousand troops. I expect that this was done not only to entertain him but to put the wind up him. It was only a sixth of the Indian Army that marched past him, and this information, casually given him by Lord Kitchener, set him thinking hard. It was said that his chief military adviser had assured him that his own army was the finest in the world, and it was also said that this adviser would provide another little job for the Royal Head-Tailor when the Amir reached home once more. The Amir visited many other places in India but was not assassinated until a few years later, in his own capital: the deed was done by one of his own loving subjects, perhaps some relative of the prophet's who wished to prove that the prophecy had not been so wide of the mark after all.

About a month after the Amir's visit I was sent down to the Fort in charge of a signal-station which worked in with Sikundra Taj. At the Fort there was always a soldier-caretaker who had the job of showing the winter visitors over the interesting buildings there and reeling off their history. At this time it was Lance-Corporal Gerald of our D company, and if the whole of India had been combed, no better man could have been found for the job. He knew the whole history of the place from A to Z and had a charming cultured voice for lecturing purposes. I believe that some of the female visitors were more interested in him than what they were in the buildings, and he made quite a lot of money in tips, which he spent right royally in booze. Gerald was never known to refuse any man the price of a drink, if he had it on him. He came of a good family and was an old Public School boy. He had studied for the Bar at first but soon got fed up with that and joined the Army in the hope of being granted a commission after two or three years' service. In the early days of his soldiering he had been getting a liberal allowance from his people, but this had been cut off after he had been in a few scrapes; it was these scrapes, too, which had prevented him from being recommended for a commission. After the Pekin expedition he had deserted in Northern China and paid a visit to Shanghai, where he spent a few hectic weeks masquerading as a certain Lieutenant St. Clair of the United States Army. He visited the best hotels and bars and made himself extremely popular with the naval and military officers of other Powers, and with British officers of any corps but his own. If a Royal Welch officer on a visit to Shanghai happened to enter the place where he was, he always made his excuses to the company and took his leave. After borrowing thousands of dollars from local money-lenders he decided that the place was too hot for him and hooked his passage on a boat leaving for Manila in the Philippines.

He spent some time at Manila as an ordinary visitor and then returned to China. He became Lieutenant St. Clair again at Hong-Kong, where his own company was stationed. After a few days he decided to give himself up. He had passed many men of his D company pals in the street and had heard one of them exclaim to another: "Well, if that toff is not Gerald it must be his bleeding twin brother!"

One evening he strolled into one of the principal hotels and was hailed by a British naval officer whom he had met at Shanghai. After they had had a drink or two a senior officer of the Royal Welch came into the bar and greeted the naval officer, who introduced Gerald to him. The Royal Welch officer said: "Do you know, Mr. St. Clair, I am convinced that we have met somewhere before, but I can't for the life of me say where."

"I am damned sure I can," Gerald answered at once. "I have been in front of you several times."

But he still couldn't grasp it. "In front of me, Mr. St. Clair?"

"At the Orderly Room, for a sequence of petty crimes, sir. I am Gerald of D Company. I have been a deserter for some time now, and I am bored to death with it, and you will be doing me a favour if you send for an escort at once. I am simply dying to rejoin the Battalion."

It caused a great sensation in the hotel when an escort of the Royal Welch arrived and marched off this dashing Lieutenant of the United States Army to the Regimental Guardroom. Gerald was tried for desertion by a District Court-Martial, found guilty and sentenced to six months' imprisonment. He had to continue soldiering after his release, and lost all his former service, which he could only regain by going for three, or it may have been four, years without an entry in his regimental conduct-sheet. But for this he would have been time-expired by now.

The second year we were at Chakrata Gerald went on a month's leave to Bombay, and a young lieutenant of the Battalion who had won the D.S.O. in South Africa happened to do the same. The young lieutenant, who was very poor, was like a fish out of water among the other officers of the Battalion, who were all wealthy men. He once told me – it shows how hard-pressed he must have felt to have made a confidence of this sort to a private soldier – that his people had a difficult job to make him an allowance of two hundred a year, and that with my pay and allowances and everything found I was really better off than what he was. Gerald and the Lieutenant met at Bombay, but Gerald would never say whether the meeting was prearranged or not. At any rate Gerald had somehow found out that a certain Captain Smith of another regiment had taken a month's leave somewhere in India; so, risking whether this officer was in Bombay or not, he decided to adopt his name and rank for the time being. He and the Lieutenant borrowed many thousands of rupees each from the Parsec money-lenders and had a delightful time together.

They had not been in the city more than about ten days when Gerald was arrested on the suspicion of being an impostor. He pretended to be highly indignant and so did the Lieutenant, who stuck by him loyally. They both told the officer in command of the Military Police that if he was responsible for this arrest he would be very sorry for it before he was many hours older. A telegram was sent to Captain Smith's battalion, enquiring whether he was on leave and giving a description of Gerald. A reply came back that he was on leave, but where exactly was not known, and that the description given fitted him accurately. This was a bit of luck for Gerald, who was immediately released with handsome apologies, including one in writing' from the Chief of Military Police in the Bombay Government. Fortified by this, he now had the daring to write to the Governor of Bombay demanding that the persons responsible for his arrest should be severely reprimanded. The Governor replied that enquiries would be made without delay. Gerald's luck held for a further ten days, but then he was arrested on the same charge. Captain Smith's colonel had written to say that he had heard from the genuine Captain Smith, who was spending his leave two thousand miles from Bombay.

Gerald returned to Chakrata under escort and when the police report arrived from Bombay he was put before a District Court-Martial. He spent over a month in the Guard-room before his trial came off, and while he was there it appeared in Battalion Orders that the Lieutenant had resigned his commission. It was common talk that the Lieutenant had been given the choice between a Court-Martial for conduct unbefitting an officer of His Majesty's Forces, and resignation. If he had chosen a Court-Martial it would have brought disgrace on the Regiment and degradation on himself, so he had no real choice in the matter. When a man is tried by a District Court-Martial he is usually defended by one of his own officers acting in the capacity of lawyer. Gerald dispensed with this assistance and conducted his own defence. He pleaded not guilty. He knew that he did not stand a ghost of a chance of acquittal, but was determined to make it as awkward as he could for the members of the Court. He astonished them with his knowledge of the Law, both civil and military – his early studies came in useful here – and embarrassed them greatly by insisting that since his chief witness, the Lieutenant, was not present they should, not only in justice but in Law, postpone the trial until he appeared. Gerald knew, as well as the Court did, that the Lieutenant was in a ship on his way back to England; and he also knew that every effort would be made to avoid recording the Lieutenant's testimony. So he was tried without this witness, in spite of his continual protests, and then had to return to the Guard-room and await the findings of the Court. His pounding away at the point of the absent witness was very cunning; he had really been calling attention to a point of honour to which the Court, as officers and gentlemen could not shut their eyes. They would be ashamed to give him the severe sentence which, this being his second offence, he had a right to expect, seeing that the officer who had aided and abetted him was allowed to slide away without a stain on his character. It turned out as he had hoped: the sentence was extremely lenient. He was found guilty of masquerading as an officer but sentenced to only forty-two days imprisonment, the sentence to run from the day he had been ordered to appear before the District Court-Martial. His sentence was read out on the very day that this period expired, which meant he was immediately released.

These were not the only scrapes he had been in and he had newspaper cuttings relating to every one of them, which he laughingly said were among his most treasured possessions. The amount of drinks he could shift without it taking any effect on him was an eye-opener. I could shift a drop myself, but he could drink me blind. In spite of his feats with beer and whiskey, he had found time to win two certificates in Hindoostani and one in Persian, and he had twice appeared in the Civil Court at Agra and successfully defended men of our Battalion who were on trial for beating up natives outside the Suddar Bazaar. At the Fort he had a bunk of his own, which was on the top storey of one of the buildings. Just outside his bunk was a ladder up to the roof, where we had our Signal Station. On the afternoon that I arrived at the Fort, Gerald had just conducted some American visitors around the place and they had been so delighted with him that they presented him with a hundred rupees. He and I had enjoyed many a drink together before, but over a bottle or two of Indian Pale Ale that evening he proposed that we should have a real solid unlimited booze in his bunk the next day. I agreed cheerfully and promised to call at his bunk about mid-day.

The Signalling Officer visited us the following morning and went off again at about half-past eleven. I then told the other three men in the Station that I was going to spend the rest of the day with Gerald and would probably not be back until late that evening. "All right, Dick, we can manage," said a man named Salisbury. When I entered Gerald's bunk I found an excellent lunch laid for the both of us; he said that it was far better to begin drinking on a full stomach than on an empty one. I agreed with him, and we washed our lunch down with a preliminary bottle or two of Indian Pale Ale. He then produced a dozen more bottles of the same brand, three bottles of whiskey, and half-a-dozen bottles of soda-water, saying if all this did not prove sufficient he could easily get more from the Sergeants' Mess. As luck had it, no visitors arrived that afternoon, and we were able to smoke, drink and yarn without interruption until very late in the evening. Afterwards I could remember everything up to the moment of leaving him: how he tried to persuade me to remain in the bunk for the night, and how I swore that it was my duty as the man in charge of the station to get up on the roof to see that the lamp was working all right, and how he argued that this was unnecessary, because the Station had packed up hours ago, and how I insisted on going. The last thing I remember was being helped up the ladder by him. After the first few rungs I told him that I could manage by myself. We wished each other good night and then there came a blank until I awoke at daybreak with a terribly fat head. I stumbled down to the washhouse and held my head under a tap of running water for about fifteen minutes. This did me good, and by the time I was back on the roof the other three were up and folding their blankets. Salisbury asked me how I felt and whether I remembered what I had done after arriving on the roof shortly after midnight.

"I feel rotten, Sol," I said, "and I can't remember even reaching the roof: I can only remember as far as about the tenth rung."

"I don't suppose your head is as bad as mine, or your body as sore," he complained. "Yes, you got on the roof all right at some time after eleven. The three of us were asleep but you soon woke us up. You wanted to know the reason why the hell the bloody lamp was not working, the same as we'd promised. You were mighty drunk, but not quarrelsome, and when I told you that the Station had closed down hours ago that seemed to satisfy you. I did my best to get you to lie down, but you wouldn't. You sang your favourite verse of the *Vicar of Bray* about 'Damned are those that dare resist or touch the Lord's anointed' in such a

213

loud voice that I'm sure old Shah Jehan and Mumtaz his wife must have groaned from their tombs in the Taj Mahal. If you had sung another bar the Sergeant of the Guard would have been in duty bound to send a file of men to rush you to the clink."

I laughed. "I don't remember that," I said.

"That was only the start of your little games. You then shouted out that you were going to do a wild Himalayan dance on the edge of the ledge of the old khudside and, before we could prevent you, you had sprung on the parapet. Dick, they say that the One Above looks after all drunken men, and, by Christ, I believe it, after last night. For a few minutes, which seemed hours to us, His entire time must have been taken up in looking after you to the neglect of all other duties. We were afraid to make a grab at you for fear you might fall into the courtyard below by trying to avoid us. Our hearts were in our mouths as we watched you perfom your puja, as you called it."

"You're kidding me, Sol," I said.

"Kidding you, hell!" said Salisbury. "You did the most fantastic steps and high kicks, sometimes dancing to the right and sometimes to the left of us. Our only hope was that when you did fall it would be in the right direction; and it was. You attempted a pirouette and down you came, ack over bloody tock. I caught you in my arms but your weight was too much for me and we both crashed on the verandah. I had all the wind knocked out of me and got a nasty crack on the head. Here, look at it, if you think I'm kidding you! You hardly stirred after you fell and the other two soon straightened you out and threw a couple of blankets over you where you lay. You're a lucky man, Dick, but if you go for a day's booze with Gerald again while we are here, I'll lie in wait for you at the head of the ladder that evening with a club, and I'll knock you insensible the moment you step off it. I don't enjoy having my feelings played upon, like what they were last night."

The verandah and parapet on the roof were like those on the double-decker bungalow. When I looked over the parapet, down into the courtyard some sixty feet below, I drew back quickly and swore that never again would I get drunk on an upper storey of a high building. Salisbury had a decent lump on his head and had undoubtedly saved me from a nasty fall, for which I was very grateful to him. (Sol was with me in the Battalion during the War and won the Distinguished Conduct Medal; he was killed in a bombardment in Fricourt Wood on the Somme, one dark night in 1916 when we were sleeping a few yards from each other.)

I ate a decent breakfast, considering the state of my head, and afterwards slipped down to Gerald's bunk. He said that he felt fine, and what about another day's booze? I told him that yesterday had been quite enough for me. All that was left of the drink supply was about a noggin of whiskey in one of the bottles, and three bottles of soda.

The following winter Gerald went home, time-expired, and the last news I had of him was a few years later when I had left the Army myself. I heard that he had called one evening at Fort Brockhurst, Gosport, where time-expired men were sent after leaving India. Here they were fitted out with civilian clothes before leaving for their homes. He enquired if there were any time-expired men of the Second Battalion there and it happened that only that morning about fifty of them had arrived at the Fort, many of whom knew him quite well. He invited them all to come along with him to the Canteen. There were not many who refused the invitation and he paid for all the beer they could drink, up to stop-tap. He also paid for all the bread and cheese that they could eat, and the pickles too; and before leaving told them that he was sailing from Southampton the following morning to take up an important appointment in South Africa. He was a remarkable man in many ways and a perfect gentleman: he never let a comrade down and gamely took whatever medicine was coming to him.

THE LECTURER

An Act came into force after I arrived in India that the terms of enlistment would be three years with the Colours and nine on the Reserve. The main object of this Act was, I think, to create a strong reserve in readiness for the Great War, which already seemed bound to break out sooner or later. Quite a number of these three-year men, as they were called, were sent to India and, because of this, did an additional six or twelve months with the Colours, according to the date of their enlistment. Many of those who joined us, however, had served two years at home before proceeding overseas, so that a man might come out one trooping season and go home the next. I was told that the Indian Government did not consider that a soldier had paid his way unless he had served three clear years in India, and strongly objected to these short-service men being sent out. I don't remember any more of these men joining the Battalion after 1905 or 1906. Probably the three years' scheme was abolished. The last large draft that we had at Agra were men who had enlisted for nine years with the Colours and three on the Reserve, but they did not have to serve an extra year at the end of the nine. I believe that this scheme remained in force for about twelve months before a return was made to seven with the Colours and five on the Reserve, which added up, as before, to thirteen years.

During the winter one of the big bugs of the Army Temperance Association visited the Battalion and presented medals to those bun-punchers that were entitled to them. During the evening of his visit he gave a lecture on the evils of strong drink, at the small theatre in the Suddar Bazaar. Everybody was given an hearty invitation to the theatre and as the Prayer-wallah and I were stony broke we attended the lecture to take our minds off the Canteen.

The lecturer was very eloquent and I enjoyed listening to him, not for what he was saying but rather for the way he said it. But one thing stuck in my throat and that was when he told us that statistics proved that there was a far higher percentage of deaths among alcoholic drinkers than among teetotallers in India. It was quite the reverse in the Royal Welch, where the teetotallers were in a great minority, and had lost more men by far than the beer-wallahs. I wanted to jump up and make this objection, but the lecturer had the best part of his listeners so hypnotized by his words that, if I had done so, I have no doubt that the Prayer-wallah was right in saying that I would have been torn to pieces by wild bun-punchers and my remains rushed along to the Guard-room for creating a disturbance. The Prayer-wallah saw me flush up with annoyance and nudged me to come away; so we slipped out quietly. The Prayer-wallah was unusually silent on our way back to Barracks, but I now knew him so well that it was clear to me that he was deep in thought over something. Suddenly he said: "Well, Dick, I have been deep in thought over that char-wallah and I have arrived at the opinion that it's a case of 'House not Correct.' The man's a damned hypocrite. Did you notice the way his hair shone, like the hide of one of those big dray-horses that pull the purge-wagons at home? Those big horses are kept in condition for their heavy work by a daily bucket of the best. Did you notice the muscles of his neck when he was holding forth? Stone me pink, but those are the neck-muscles of a proper soaker. And did you notice the delicate rosy hue of his face? He keeps his pores clean with purge, which taken internally is the best complexion-cream known in this unhappy world. Char dries a man up and turns his face yellow – look at the Chinks, who first started the char-drinking habit! No, Dick, he's a man who if he were lost in the Great Mongolian desert would far prefer a pint bottle of Pale Ale to a whole hog's-head of pure water. What first aroused my suspicions was the face he pulled after he had taken his first sip out of that glass of water beside him. What convinced me completely was the face he pulled after his second sip. I'd wager my life that the lecturer could drink you and me and Gerald silly and incapable, and never turn a hair."

I said: "No, I don't believe a word of it. It may be that he was a beer-wallah once, but in my opinion he's been converted and some of the strength that purge-shifting gave him has gone into his speech-making. That's the pity of it."

The Prayer-wallah said: "Once a purge-shifter, always a bloody purge-shifter, unless a man's forced to go on the tact. But the wages he'll be earning as a lecturer must be making a rich man of him. He'll have a fine soak tonight. If you and I had only a tenth of the neck-oil he's going to shift this evening we'd sleep a hell of a lot better than what we will do."

I argued the toss; being on the tact put me in a argumentative mood. But the Prayer-wallah said: "You Taffies are all the same. Any man with a fine eloquent speaking voice can take you in, the same as if you were children. That's why religion goes so strong in Wales and that's what gave Lloyd George his start in life."

"You'll be telling me next that that damned bun-punching char-wallah has converted me," I said, a bit angrily.

"Well, I wouldn't be surprised," he replied.

We walked the rest of the way in silence. But the story does not end there.

During this summer of 1907 the whole of the Signallers were sent to Kilana. After being there for about three months, four of us were ordered to proceed to Landaur to carry out some experimental work with a nine-inch heliograph and a "CC" lamp. Our orders were to work for a fortnight in conjunction with our signallers at Kilana, who would be using the same instruments. After this we would have to do whatever work was found for us at Landaur. These large signalling instruments, which were not issued to units, were kept at the depot stores at Hill-stations. Both had strongly made wooden cases which were too heavy to be carried any distance by manpower; but they were so made that they fitted one on each side of a mule-saddle. Taking two days' rations with us we set out on our journey accompanied by two mules with their drivers, the mules carrying our kits and blankets. The track we travelled was very rough in places and after making a gradual descent of fifteen miles we arrived at a small stone dak-bungalow. After leaving Kilana a mile behind us we had not seen a solitary human being until we reached this place, where two natives were employed as caretakers. Their job was a lonely one, as there were no signs of any life anywhere around this desolate looking spot. This bungalow was only a resting-place for travellers to put up for the night, and no food could be bought there; only bottle-beer, but at two annas a bottle dearer than at Kilana. When we grumbled, the caretakers said that it cost much pice to hire a hill-coolie to bring cases of beer from Kilana and take the empties back. But the beer was well worth the extra charge and just before dusk the four of us were seated outside the place drinking, smoking and yarning. It was a scene that I shall never forget, because as dusk fell the distant hills in front of us towards the Snowy Range, which we now could not see, were lit up by a long chain of fires. It was a very beautiful sight, but a bit alarming because the frontier tribes sometimes used beacon-fires as a signal for a raid or a general rising. We asked one of the mule-drivers what was the object of the fires and he explained that the people living around those parts were fire-worshippers, and that they were holding a big Ramsamee on this night. The natives in the bungalow said the same and by straining our eyes we could dimly make out shadows dancing around the fires, which were still burning when we retired for the night. The following morning we made a sharp descent of about a mile into a valley where we crossed a small bridge over a little mountain stream. The heat was stifling in the valley and a notice-board by the side of the bridge told us that although still in the heart of the mountains we were now only a few feet above sea level. After crossing the bridge we made a gradual ascent for fifteen miles until we arrived at Landaur.

Here we were attached to a company of the Royal Irish Rifles. We were provided with a small room on our own and got on first-rate with the men of this battalion, who were all from the North of Ireland. Their N.C.O.s never interfered with us in any way. Our signalling gear was kept in a small store close to the spot where we worked in with Kilana. A signalling officer of the 17th Lancers was in charge of us and sometimes spent an hour with us in the morning. After we had been working a week the rainy season began. I remember how one clear morning, after we had fixed our helio and telescope and got into communication with Kilana, the fog crept slowly up the hill until we could not see ten yards beyond us or behind us. And yet the sun was shining dully through the fog, and the rays of those large heliographs were so powerful that with the aid of the telescope we continued to exchange messages until the sun had totally disappeared. But if, by some accident, we had accidentally dislodged the legs of the heliograph or telescope stands all signalling would have ceased: it would have been impossible to reset them on the distant station until the fog had lifted. On a clear day this heliograph could be easily read at a distance of over a hundred miles; the distance that a "CC" lamp could be read, according to the Signalling Manual, was twenty miles with the naked eye, and forty with the telescope. With good visibility it could be read much further than this: on very clear nights we sometimes read messages with the naked eye from Kilana, which was thirty-one miles away as the crow flies, but the rays of the lamp would not penetrate a fog.

After we had finished our work with Kilana our Signalling Officer informed us that he would let us know in a day or two what work we would have to do next. He was under the impression that when the rainy season was over we would work our heliograph in conjunction with Roorkee, on the Plains sixty-six miles away, which was the headquarters of the Bengal Sappers and Miners. We never saw him again. I believe he either went on shooting leave or rejoined his regiment on the Plains. For over two months, with the exception of our rations and pay, which were the main things that mattered, we seemed to have been entirely forgotten. When we felt like it we did a little signalling work among ourselves, but most of our time was spent in exploring the neighbourhood. Opposite Landaur is a well-known hill called Fairy Hill. I was told that several attempts had been made to build bungalows on it but there was so much iron ore in the ground that they had always been struck by lightning before they were completed. A few miles from this hill were waterfalls and sulphur springs. One of the falls took a leap of about two hundred feet and then rushed over a huge rock into the stream below. Mussoori was a fashionable Hill-station where the rickshaw coolies trotted the toffs around. I was surprised at the large number of churches, convents and schools at this station, but not being interested in such places I did not investigate them closely.

At Landaur was a very large Soldiers' Home, run by a man called Jimmy Taylor, a North Country man. There was a similar one at Meerut, but I don't know whether they were Jimmy Taylor's property or whether he was only managing them. They did an excellent trade in the evening selling cooked suppers to the troops. It was customary there to hold evening prayer-meetings in one of the rooms, where the so-called Bible-punchers sang hymns until they were black in the face. There were not many genuine Bible-punchers in the Royal Welch. I can only remember about a half dozen of them during the whole of my soldiering days. One of them that I knew very well was always doing his best to convert others and had all the usual Salvation Army patter on the tip of his tongue. I thought he was very genuine until I met him one night sneaking out of a shack in the brothel at Agra. I asked him whether he had succeeded in converting the girl he had been with, plucking her out like a brand from the burning; but he did not reply and walked away with a scowl on his face. Jimmy Taylor was very popular and I attended many fine concerts that were held at his Home, which was so large that he was able to accommodate a decent number of men who were on a month's leave from the Plains. He only charged them a rupee a day for their food, which they all agreed was first class, and for their sleeping accommodation; the reason why he could do things so cheaply was that by arrangement with the military authorities he was allowed to draw the men's rations. During the summer men could have a month's furlough to go anywhere they liked; during the winter it was very rare that men were granted furloughs in the country, but a man could always have a few days' leave to attend, say, a race-meeting, sports, or a boxing-tournament held at some station away from his battalion. Men who went on furlough were attached for rations and sleeping accommodation to the units stationed at their destination. Of course, their time was their own, but they came under the discipline of the units they were attached to. Officers, I believe, were allowed ninety days' leave in the year. After every three years they were also allowed twelve months' leave at home on private affairs.

One evening I and two others were taking a stroll just before dusk down a zig-zag mountain track leading to Mussoori. We had not gone very far when we heard a distant noise of singing. It was a very powerful voice but the echoes at first confused the tune, until when we turned a sharp bend it burst full on us. "Put me among the girls, Bright eyes and golden curls." One man said to me: "That will be a colour-sergeant or a senior sergeant at least by the voice. It's got the parade-ground ring to it." "Coming back from Mussoori with a skinful," said the other. But out of the dusk emerged a group of three men: two natives sweating along up the track under the weight of a high sedan chair and a big civilian lolling in it; mighty drunk and carolling away above their heads as if he had not a care in the world. He put me in mind of an illustration to Plutarch's *Lives* – a drunken Roman Emperor carried through the streets of Rome by his slaves in a similar conveyance. The track, which led to the private bungalows, was not much used by the troops, so it must have been a bit of a shock to the lecturing char-wallah (for it was he) when he saw us. He pulled himself together and said in a thick voice: "Good evening, boys! Beautiful stars above us, beautiful night, eh?" We agreed, having a difficult job not to laugh outright. As soon as he passed out of sight around the bend he broke out again with: "When I marry Ameliah, won't we have a go? Clocks with ruby faces, Emerald dressing-cases ..." The Prayer-wallah had been right.

Towards the latter end of October somebody at Brigade Headquarters at Mussoori, whose orders we were under, discovered that we were still at Landaur. The Lance-corporal in charge of us was informed by the Adjutant of the Royal Irish Rifles that our experimental work had finished and that we would proceed on the following day to rejoin our battalion at Agra. We paraded at Orderly-room shortly after noon the next day, and, because we had been working so extremely hard all this time, four mountain-ponies, called tats, were provided to convey us on the first part of our journey. Two mules were also provided to carry our kits and blankets. Our orders were to proceed to the Rajpore Hotel, but fourteen miles distant at the foot of the hills. On arrival there we were to hand the ponies over to the Hotel, from where a mail tonga would take us to Dehra Dun, from where we would travel second-class by train to Agra. Officers of the Viceroy's staff could hardly have had better travelling arrangements made for them than what we had. At Agra, the Sergeant-major, to whom we had to report, did not know that we had been at Landaur until we told him. We allowed him to believe that we had been doing a gruelling three-months' work, so he told us to attach ourselves to the Telegraphists and take things easy until our own chaps arrived from Kilana. Another week passed away before they did arrive and then our long holiday came to an end.

I was glad to see the Prayer-wallah again. For over six months he had been strictly on the tact: in fact, he had been living on the skin and saving every pice of his pay. He was due to leave the country time-expired in January and, as he explained, he wanted a few corks to land in Blighty with. A day or two after arriving from Kilana he made me a present of his revolver, with its remaining ammunition, and his magnifying-glass, saying that the glass had been of far more service to him than the gun and that he hoped that it would stand me in good stead in Burmah, where the Battalion was now under orders to proceed. I was honest enough to tell the Prayer-wallah that he had been right and I had been wrong about the lecturer. He laughed and said: "Well, if that bun-punching old bastard was as drunk as you say, he must have soaked enough of neck-oil to have put a whole company of men clean out of action. What I don't understand is, why doesn't he lecture on behalf of the big brewing companies on the virtues of alcohol? He ought to be able to pick up good money that way, and I should greatly enjoy listening to him." But I said that purge was its own best lecturer and that the brewers didn't need to waste their money on paying a man to hold forth in a lecture-hall about what could be learned far better and more persuasively in the Canteen, every man for himself.

A week later the Prayer-wallah complained of being feverish. After hanging on for a few days he felt worse and finally had to fall sick. Before going to hospital he handed me his belt with all his savings in it, saying that if anything happened to him I was to keep it and have a good drink in memory of him. I told him not to talk so damned wet and that I was surprised at him getting the wind up. "No," he replied, "I haven't got the wind up, but some strong healthy men have been unlucky enough to pass off this Ball of Clay in double-quick time since we have been at this station. For all I know I may be one of the unlucky ones and I tell you, Dick, I am iller than what you think. I should never have gone on the skin." He was admitted to hospital and found to have enteric fever. With the assistance of the orderlies, when the nurses were not about, I paid him occasional visits in the evenings, until he got too far gone to recognize me. The orderlies said he was a grave case but believed that he would pull through.

It was touch and go with him until after the twenty-first day of his illness, when to my great relief he took a turn for the better.

I had felt very low during those three weeks, as can well be imagined. I visited him during the evening of the next day and handed him his belt. He wanted me to take a twenty-rupee note out of it to celebrate his turn for the better, but I refused, saying that he would want every clink he had when he returned to Blighty. This was the last time I saw him, for we Signallers, and the Band, left the following day for Calcutta, our port of departure for Burmah; we were being sent on three days in advance of the Battalion. The Prayer-wallah came from one of the Midland towns, but had been working in Lancashire when he enlisted. He was the only real pal I had in India, and there is nothing that we would not have done for each other, but the strange thing is that I have never heard a word from him or about him from that day to this. When I was called up as a reservist at the outbreak of the Great War I made inquiries about him with men who were time-expired at the same time as him, but not one of them could remember whether he was invalided home and not one of them had ever met or heard talk of him in civil life. If he is still alive and kicking somewhere it would do me a lot of good to see him once more.

Our Regimental soccer team had met with a fair amount of success since the Battalion had arrived in India. They had won two or three Cups open to the troops in Northern India and had only been beaten by a solitary goal in the final of the Simla Cup in 1907. The Cameronians were the victors. The most amusing soccer match I ever saw was played on a wide open rough road at one end of the Regimental Bazaar. The players were two picked teams of natives from the Bazaar, who played in their bare feet. Their ages ranged from ten to seventy. They had a proper soccer ball, but the goal-posts were represented by large blocks of baked clay. The chief difficulty of the native referee was to judge whether a high ball which had passed over the heads of the goalkeepers was a goal or not a goal. Most of the male population of the Bazaar had turned out to watch the game and as it progressed they became as excited as the crowd at a Cup Final at Wembley.

Most of them had wagered a pice or two on the game, and by the row they kicked up anyone would have thought that there were thousands of pounds in the pot. I heard more bad language used in this match than what there would have been in a dozen matches between the Royal Welch Fusiliers and the Highland Light Infantry. Young and old cursed one another with impunity and the referee spent a great deal of his time cursing players and spectators who had dared to question his decisions. Just before half-time the first real stoppage in the game occurred, when the referee awarded a goal for a ball that had passed a good fifteen feet above the head of the white-whiskered old goalkeeper. The goalkeeper called him a something-or-other loose-wallah, and the game was held up to allow these two to stand with arms folded on their chests at the regulation interval and go back in each other's pedigrees. The duel lasted about ten minutes, and the referee, who had a bit of a reputation for this sort of thing, which was his chief claim to the position of referee, was declared the winner. Not many players or spectators objected to his decisions afterwards. Halfway through the second half, a large Indian cart drawn by two slow-moving bullocks came along the road and passed over the playing pitch. The driver and his ancestors were cursed to all eternity by the referee, the players, and the spectators, for interrupting the game. The driver halted his team in order to reply suitably, which lengthened the interruption considerably. Just before dusk the game ended up in a draw of 1-1.

I cannot remember any natives playing rugby, but they were enthusiastic soccerites. When we passed through Roorkee on the line of march from Meerut our Regimental team generally played a game with an Eleven of native students from the college of this place. The students, who played in their bare feet, were a fast scientific side and their full backs were powerful kickers. But one thing that they could not stand was an honest shoulder-charge, which was the chief reason why they were always defeated when they played our chaps.

It has been said that wherever a white man goes he takes with him, accidentally, in the corn he has with him, the seed of the Buttercup, which springs up around his settlements; so that the Buttercup is called "The White Man's Footsteps." I don't think much of this yarn, because though it may be true of South Africa and Canada and similar places, I cannot remember seeing any specimens of that flower growing wild around the troop-stations of India or Burmah. But if one were to use the expression about the game of soccer, it would be true enough, and if we are ever so unfortunate as to lose our Empire it is a safe bet that soccer will continue to be played all over the world – in jungle-clearings, on the open veldt, in the sands of Egypt, and on hill-tops in the foot-hills of the Snowy Range – in eternal memory of the British soldiers who accidentally brought the game along in their kit-bags.

BURMAH

Colonel "Tommy" Lyle, who had completed his four years of command, left the Battalion just before we left Agra for Burmah. We were all sorry that he was leaving us; even the hard cases, who could not keep out of trouble if they tried, admitted that it was impossible to have a better commanding officer than what he had been. His successor, Lieutenant-Colonel "Paddy" Mantle, was also a popular commanding officer. As a young officer he had served with the Battalion in the Burmese War of 1885, and had won the D.S.O. for gallantry at the capture of Bhamo Fort. The journey by train to Calcutta took us several days and nights. We stayed at Fort William until the Battalion arrived, but were not allowed out in town. The Fort was a very large place but none of us had much time to look around it, as most of our time was spent unloading the Battalion's heavy baggage from the train and reloading it on the Royal Indian Marine ship that was to take us across the Bay of Bengal to Rangoon. During the evenings, for the three days we were here, I spent my time in the Canteen. After leaving it, I stayed up late in the night playing Pontoon, a card game which also goes by the names of "Twenty-One" and "Van John." The Goddess of Luck was more than kind to me. I won over four hundred rupees and this money enabled me to have a proper beano when I arrived at Rangoon, which is a place well suited for beanos. Only a certain number of dogs belonging to the Battalion were allowed in the ship, but in some mysterious way every owner smuggled his pet aboard. The man who owned the cross between bulldog and greyhound said that this creature was more important to him that any officer's bloody polo-pony.

There were no hammocks aboard the ship, so we were issued

with blankets. We had not been many hours at sea when orders were issued that no man was to sleep on the upper decks. This was rather tough, as we found on laying our blankets down for the first night that the sleeping-accommodation below deck was altogether insufficient for us. The Master Gunner, who had been a first-class warrant-officer in the Royal Navy, was in charge of ship's rounds which came along the decks before lights out. The first night he came along the deck I was on, not much gangway had been left for him, which caused him to swear a bit. A corporal pointed out that our quarters were so cramped that there was no room to leave a decent gangway for him. He roared back at the corporal: "Room, Room! What the hell do you want? A bloody floating hotel to sleep in? Dammit, man, there's enough of bleeding room on this deck to accommodate the whole of the crews of the combined Mediterranean and Atlantic Fleets. Why in Christ's name don't you sleep like sailors do, fore and aft? Try it tomorrow night and you'll be bloody well surprised at the great wide empty spaces left over." Something had upset his apple cart before he came around our deck, for he did nothing but curse and swear until he had left it. His last words were that a gouty old Port-Admiral taking a cruise for his bleeding health would have been ashamed to have taken up the sleeping room that each one of us was taking up – sprawling about like the Empress's favourite kitten in the great Imperial nuptial couch of Hell-knows-where. The majority of the men were sleeping fore and aft, as a matter of fact, but he pretended not to notice them.

The following night, Salisbury and I, who had left it a little late to lay our blankets down, found it impossible to work in edgeways between any of the men. We were standing with our blankets under our arms at the foot of the hatchway when the Master Gunner came down it. He was in a more cheerful mood than on the previous night, but I thought he was going to explode when I told him that the only possible place we could lay our blankets upon was the wide gangway that had now been cleared for him. Before he could think of a suitable reply I politely asked him whether he would permit the both of us to sleep on the upper deck.

He did a huge guffaw at this and said: "Well, I go to bloody Boston! But, in the words of Scripture, 'Ask and it shall be given unto you.' For your God-damned blasted cheek you two birds can roost up there tonight, and tomorrow night too, if you wish. And if anybody questions you, just tell them that the Master Gunner has given you permission."

We saluted him gratefully, though he was not entitled to a salute really, and moved off. As soon as we arrived on the upper deck we were unlucky enough to be spotted by our Regimental Sergeant-Major, who was a pompous individual. In his best Sergeant-Major manner he demanded an explanation of our appearance on the upper deck. I gave it. He then said that he wouldn't care a damn if the ship's captain himself had given us permission. We were not going to sleep there: orders were orders, and if we did not get below in double-quick time he would make the damned pair of us prisoners. Down we trailed again with our blankets and waited at the bottom of the hatchway until the Master Gunner returned.

"What? Isn't the fo'c'sle good enough for you two lobsters?" he asked.

I told him what had occurred to us. He commenced to swear, saying that no blasted two-bit Sergeant-Major was going to overrule his orders. The fact was, that the Navy being the Senior Service, Naval ranks and ratings took precedence in seniority over their opposite numbers in the Army, and Master Gunner was a very high position. He told us to follow him to the upper deck, where we again met the Sergeant-Major, who asked him if he had given us permission to sleep on the forecastle.

"Yes, I did, and what the bloody Hell has that got to do with you?"

"Everything," replied the Sergeant-Major. "The Commanding Officer's orders are that no men are allowed to sleep on the upper deck; and that is the reason I sent them below."

The Master Gunner now poked out his chin and roared: "Who in Christ's name are you, anyway, and for the matter of that who the effing blazes is your Commanding Officer? Remember, my man, the Captain is in command of this ship. He has power of life and death over every man Jack in it. He has the power to flog and the power to hang, not to mention the power to dock your booze and clamp you in irons, and the power, if he chooses, to shave off those dainty waxed moustaches of yours close to your lip. The Captain is God Almighty in his ship on the High Seas. He can register a bloody birth, solemnize holy bloody matrimony and sign a man's bloody death certificate; and he can make Easter Sunday fall on Good Friday if it so pleases him. And, what's more," he said, shoving his chin forward until it was within a couple of inches of the R.S.M.'s face, "I'm the Captain's bloody second dickey. I have ordered these men to sleep on the fo'c'sle, and sleep there they bleeding well shall – if you dare to order them below again you just see what happens to you. And I promise you, Soldier, it will be a frightful fate."

Being of inferior rank the Sergeant-Major was unable to talk back at the Master Gunner, so he wisely made himself scarce. We slept on the forecastle without being disturbed, and the following night as well, which was the last night of the voyage.

On arrival at Rangoon the Battalion encamped for a week at the back of the Shway Dagon Pagoda, or Golden Pagoda as it is called, supposed to be the finest in the world. It has a spire which is encased in solid gold leaf. Every Buddhist pilgrim who visits the Pagoda is anxious to attain merit by giving some gold leaf to add to the spire, which is a familiar landmark for miles around. Somewhere in the top of the Pagoda is a chime of delicately made silver bells. During the stillness of the night they can be heard tinkling sweetly and melodiously, and seem to lull one off into a gentle sleep. I visited the Pagoda. At the main entrance, one on each side, were two gigantic carved wooden statues – half man, half beast. Inside the building were other grotesquely carved figures; but everything I saw was too misshapen and far-fetched for my taste.

Close to the camp was a large lake where most of us indulged in a swim; late in the afternoon on the first day of our arrival, one man who stayed in the water after the others had packed up got drowned, and his body was not recovered until some hours later. We were allowed out in town but I did not go out myself the first evening. Half the Battalion did, and they were loud in their praises of it the following day. They said that it was the finest place in the world, and that it was impossible for a man to go wrong, either in his choice of a woman or of a bottle of beer. Salisbury, I and a man called "The Benny King" set out on the afternoon of the second day for a beano. We were strongly recommended by some of the men to take a stroll down Twenty-Nine street before we went anywhere else. The houses in this street were about three storeys high and living in them were many hundreds of Japanese girls. Most of them had been bought cheaply in Japan and brought to Rangoon for the purpose of making money for their owners. They had the reputation of being the cleanest girls in town at their profession and each one of them had a little room of her own. Most of them had not yet reached the age of twenty. The finest steaks and chops I ever ate abroad was in an eating-house run by a Japanese at the end of this street. Bottled beers from every country in the world could also be bought at this house, Bass being sold in quart bottles. There was another street full of girls of different nationalities of the East; and in another street, on their own, were a large number of European girls. Somehow I felt thankful when I was told there was not an English girl amongst them. There were also gaming houses in the town which were open day and night. For a soldier who had money to spend, it was a glorious place. The three of us had a magnificent beano that day and on each succeeding day we were there. There was not much we did not see and not much we did not taste.

Late one evening we came across a vast crowd of Burmese, Japanese and Chinese watching an open air performance of a marionette show; we watched it ourselves for about an hour. It was far more elaborate than the Punch and Judy shows at home, which are only intended for children, and was given outside a large erected wooden building draped with a sort of native canvas. On a long balcony, about twelve or fifteen feet above the ground, the dolls appeared and retired like genuine actors. They were dressed in coloured silks and were about four or five feet tall. A blood-curdling melodrama was acted in the Burmese language, the dolls being made to speak by people behind the curtain, and it ended with the villain's head being cut off by the hero at one stroke with a long curved sword. The day before we left Rangoon for Upper Burmah the three of us, late in the afternoon, visited a temple to see a pool of sacred snakes which was attached to it. It was a pretty large pool with snakes of all sizes in it. Some of them must have been twelve feet in length and I was told that they were fed daily by the priests. Although we had been drinking rather heavily Salisbury and I had also been eating heartily as well. The Benny King had scarcely eaten anything and it was either this one-sidedness of his enjoyment or the effect of seeing the pool of sacred snakes which sent him "in the rats" as we called it. For over a month he was in a pretty bad state: many nights before getting into bed he could see snakes crawling all over it. "Look at the bastards," he would shout as he drew shuddering away from his pillow. It was very little sleep he had during this bad time of his, and it was touch and go whether he would have to be taken to hospital and put into a padded ward until he got better. He got all right in time but he never drank anything stronger than Canteen beer afterwards. He was a great reader of books and also a very clever sketcher, but during the winter, whether he had taken any beer or not, he never failed to have a benny, which was the reason of his nickname.

The wives of the officers, the married crocks, and the advance party of the Battalion, had left by train for Shwebo. All passengers for Shwebo left the train at Mandalay and crossed the river on a ferry boat to a place called Amphora Shore, where they embarked on another train for Shwebo, which was as far as the railway ran in Upper Burmah. The Battalion was making the journey by river to a place we afterwards called Chum Yum, which was fourteen miles from Shwebo and about six hundred miles from Rangoon. We commenced our journey on Christmas Eve on two river-flats lashed one on each side of a paddle-screw steamer which had a Burmese pilot aboard. Owing to the sandbanks in the river it was impossible to travel by night and our average speed was about fifty miles a day. We started off each morning at sunrise and anchored at sunset, we were then taken ashore for a half hour's march and returned to the flats for the night. We would have much preferred sleeping on shore, as there was hardly room on the flats to wink an eye in comfort. The officers' chargers and polo-ponies which were aboard really had more room than what we had. The Irrawaddy is broad in parts but so narrow in others that the branches of trees on the banks brushed over the flats as we steered through. After passing the main towns of Lower Burmah, which are on the river, we met coming down the river, also on flats, the battalion of the Loyal North Lancashire Regiment whom we were relieving at Shwebo and with whom we exchanged greetings. Our rations for this trip, which included Christmas Day, were bully-beef and biscuits. By the look of the biscuits, which were full of weevils, they must have been surplus stock left over from the days of Clive, which the Government had lately discovered in some cellar and was now giving an airing. We were allowed to buy a pint of beer a day, which was brought in dixies by the section-corporals from somewhere on the steamer. It is an old saying that "there isn't no bad beer," but after trying a couple of pints I came to the conclusion that there was a catch somewhere in the saying. Beer is beer, but everyone was now crabbing about this gut-rotting stuff and arguing hotly whether it was shark's or horse's, or a blend of both with a dash of bilge to sweeten it. The dry canteen aboard sold all manner of tinned stuff, like salmon, sardines and anchovies, but I could not enjoy anything I bought eaten without bread, which could not be bought for love or money. We were all fed up by the time we arrived at Chum Yum.

This was simply a spot where troops disembarked for Shwebo and re-embarked for Bhamo Fort, which was a few days' sail up the river, near the Chinese frontier. The river at Chum Yum must have been over three-quarters of a mile in width. Five of our companies were for Shwebo, the remaining three for Bhamo. I went to Shwebo. After being cooped up on the flats, the fourteen miles' march to the barracks there seemed a flea-bite.

Our bungalows, which were a mile from the village of Shwebo, were wooden ones built on poles about twelve or fifteen feet in height. Most of the Burmese lived in bamboo huts, also erected on poles a few feet above the ground. Just behind the Regimental Bazaar were about a dozen small dirty huts and in them were a dozen or so of Burmese and Chinese girls who did not look any too clean themselves. This was the Rag for the use of the troops. Always mindful of the health of his men, one of the first things that Colonel Mantle did was to have the girls sent away from the place and the huts pulled own. Brand-new huts were erected in their places, the cost of which, I believe, came out of the Canteen funds. They did not take long to build, and it was money well spent. After they were completed, a dozen clean Japanese girls were imported from Mandalay to inhabit them. This proved to be a very wise move of the Colonel's: during the fifteen months I served in Burmah there was never a case known of a man contracting venereal with the Japs in the Rag. The few cases that were contracted were with the Burmese and Chinese sand-rats around the place.

Casualties among the troops were light, but during the first few months we lost a number of dogs. When it was discovered that the Burmese were great lovers of dog-stew, men kept a more wary eye on their pets. Only the man who owned the bull-greyhound boasted that he had no need to worry. He said that the niggers were superstitious and that this beast was too much like some of the carvings in their Pagodas for them to dare to lay a finger on him. Although the village of Shwebo was out of bounds for the troops I strolled through it dozens of times. It was not a large place, but there were a number of dog-stew shops in it which were well patronized by the Burmese and Chinese.

The natives of Upper Burmah were of a more robust build than those of India. At a very early age the males were tattooed around the legs with rings of what looked like grinning devils. This was called "the Burmese stocking" and was supposed to avert illness and enchantment. There was also a yarn going around that their backsides were tattooed with still more frightful markings; but I cannot say whether this was so. When I first saw the females I was mildly astonished at the way they wore their robes: in a manner so that when they were out walking the whole of one leg from the groin down was perfectly bare. This was a custom that was supposed to date from the same time as the tattooed backsides of the men. The yarn was that an old King of Upper Burmah, deploring the fall in the birth rate of his dominions, called his wise men together and asked them the reason for this sad state of affairs. They replied that the practice of sodomy had been introduced from China of late and gained a great hold on the country and consequently the women were being neglected. The King asked his wisest counsellor what steps to take: should he make sodomy punishable with death at the stake?

The counsellor replied that burning half his subjects wasn't going to send up the birth rate, and that the cure of a disease lay in removing its cause.

"What is the cause, in your opinion?" the King asked.

"The cause, Your Majesty, is that the women are too swathed and veiled and wrapped about and secluded to attract the young bachelors who can't yet afford to marry; so these soon fall into bad habits."

The result was an edict that the females should go unveiled and wear their robes in this highly enticing way and that the males should be tattooed in a manner to make them as uninviting to their fellow males as possible. The penalty for non-compliance with the order was death. As the result of this order the birth rate bounded up.

236

I have been told by an English police-detective that the severe penalties imposed on unnatural practices in our own country by an Act of 1886 have merely had the effect of advertising them as an interesting novelty to the public, who before did not think or trouble much about them; and that such vices are now far more widely practised than what they ever were before. This detective agreed with me that the old Burmese counsellor was infinitely more wise than our Parliament in this respect.

All the males wore long hair, which they tied in a bun at the back of their head. They were passionately fond of gambling and very lazy. They made their wives work, but did little themselves. The women were not considered of much account and could be bought cheaply as wives. I knew one Burmese man who possessed twelve. He left one in his large hut to do the cooking for the evening meal and each morning marched the other eleven to work. With flowers stuck in their hair they marched in Indian file behind him, each one smoking a Burmese cheroot, which was a daily luxury he allowed them. They generally worked in the rice-fields, or at any paid job he had found for them. The only work he did himself was about an hour's loving attention to his bun of hair. This he would unroll and well oil with cocoanut oil before tying it back in a bun again. The rest of the day he would spend yarning and smoking with other fortunate husbands that were lolling about at the edge of the rice-fields or wherever it might be. He drew his wives' wages at the end of the day and marched them back home. There the wife he had left behind would place the whole of the evening meal in front of him. His wives would then have to wait respectfully until he had filled his belly, before they could fill theirs. He would then leave them enough of money for the following day's rations and cheroots and go off for his evening gamble at the village. It never worried him or others like him if he got broke: he was always sure of a gamble the following evening, so long as he had found work for his wives. It put me in mind of an old music-hall ditty of those days, it may have been one of Dan Leno's or Albert Chevalier's:

Should husbands work? Well, I say "No!"
The *Telegraph* is barmy and to married men a foe:
Let them work as wants to work, but what I wants to know
Is: "What does a man get married for?"

I was told that the translation of Shwebo in English was "The Land of Snakes." If it had been "The Land of All Things that Creep or Crawl" it would have been much nearer the mark. I saw more snakes, black scorpions, scorpion-spiders and centipedes around these parts than any place I had been at. This, perhaps more than the fear of flooding during the rains, accounted for the huts being raised on poles. Yet our chief trouble was bugs. The wooden bungalows were infested with them. Our bed-cots, which were the same as in India, were two iron trestles on which rested two sheets of corrugated iron, each a little over six feet in length. Each man was issued with two bed-ticks, and the bed itself consisted of one of the ticks filled with coir, which resembled the hair on the shell of a cocoanut. These coir beds soon became very hard, so once a month most men changed their ticks and had the old ones picked. This job was done by poor Burmese boys who knocked around the Barracks. They charged two annas, which was one dearer than in India. But they well earned their money, because the coir, which had formed into hard balls, had to be pulled out in strands before it was put into the clean tick. It took a good bed-picker about a hour and a half to pick a single bed and stitch up the tick after the coir had been put in it. Most of the men in the companies spent a few hours on Thursday mornings bug-hunting. We Signallers did the same on Sunday mornings, which was our day of rest. The trestles and corrugated sheets were afterwards well wiped over with rags which had been dipped in paraffin; but no matter what we did we found it impossible to reduce the numbers of the enemy. It was as bad as lice in the trenches in the early days of trench warfare.

RISHIS AND FAKIRS

I was never at Bhamo, although through conversations with the men of the companies that were there I could give a fairly accurate description of the place. There was hardly a trace left of the old Burmese fort where Colonel Mantle had won his D.S.O., and the modern fort, where our three companies and a native regiment of infantry were, was about a mile from the river. There was also elephant and mule transport at Bhamo. This town was very prosperous owing to its river trade and its large trade with merchants who came with caravans from China. These caravans consisted usually of sixty to eighty sturdy ponies, all heavily laden with merchandise. They arrived by way of the Great North Road, or China Road as it was called. There were bamboo forests around Bhamo, and on each side of the China road as it ran towards the Cochin Hills were rubber trees. These were tapped for the juice which was collected by natives in pails suspended from a yoke on their shoulders.

At Shwebo the Signallers and Drummers stayed in the same bungalow, also the Goat-Major, who was always attached to the Drums. He was not the old Goat-Major of Meerut, but one who had been appointed after the arrival of the new Goat at Agra. He was a tall, dark, handsome man who took more pride over his head of hair than what a Burmese did. He was a thrifty chap, too, his only expenditure being cosmetics for his hair, and tobacco. Since we had arrived at Shwebo he had commenced to study hypnotism in an endeavour to imitate the successes of Yank. He bought a number of books on the subject, which Yank told him were not a bit of use to him. He thought that Yank was jealous of him as a rival magician and after devouring them from cover to

239

cover he began practising concentration of thought. It was very amusing to see him squatting on his box in the correct cross-legged attitude, like in fig. 1 of his illustrated manual, concentrating on some article at the foot of his bed. The article at first was either a mango, or a rolled pair of socks, or his manual itself; but afterwards it was a cheap brass figure of Buddha that he had bought for this special purpose, engaged in the same long-drawn-out contemplation as himself. The Buddha and the Goat-Major squatted opposite each other with the same intense, empty look on their faces, for the space of many hours a day.

Most of us began to think that it was only a question of time before the Goat-Major would go stone potty. He had his leg pulled unmercifully and many attempts were made to distract him by introducing queer exciting or horrible objects into other parts of his field of vision, but he seemed to take no notice of these at all. After he had been concentrating for a couple of weeks we were surprised to see him with a queer gadget fixed on the back of his head, staring at the object on his bed; which was a rolled pair of socks again, because someone had hidden his Buddha for a lark. This queer gadget, which cost thirty rupees, was an electric contrivance which was supposed to introduce gentle waves of electricity into the brain. He had seen it advertised in a Theosophical magazine, so had ordered one from a chemist at Mandalay. He called it a brain-generator. After he had used it a week, he told me that he was able to concentrate a lot better with it than what he had done before. He also said that he was making decided progress in the hypnotising line.

A week later he had smashed his brain-generator to a thousand pieces, and also a crystal gazing-ball that he had lately bought, and he had made a small bonfire of his books on hypnotism. He swore that he would half-murder the man who ever mentioned the subject of hypnotism to him again. For though the generator may have flushed his brains inside his head, it had also flushed a patch of hair off the outside of it, leaving quite a respectable

240

bald patch like the tonsure of a Roman Catholic priest. The Goat-Major now concentrated on the study of hair-restoration, and the chemist who had sold him the generator did a roaring trade in selling him bottles of hair-restorer. He used every kind of patent hair-restorer known to mankind and also the local cocoanut oil as well; but never a hair grew on that bald patch again. He was a bit sensitive over this for a while. His hypnotic studies had been a costly affair, but I believe that the generator had saved his reason, for one of the Drummers swore that on several occasions he had watched him trying to hypnotize the Goat in the Goat-shed.

As for Yank, in addition to carrying on his occult studies, he had been studying the Burmese language, which I personally could never make head or tail of. He took a few days' leave somewhere and, when he returned, told me mysteriously that he was one step nearer his goal. This summer he was away again, on a whole month's leave; I think near Mandalay, attached for rations to the East Yorkshire Regiment. Like myself he had then only eight months to serve to complete his time. When he returned he surprised me by taking on to complete his twelve years with the Colours. He told me that during this leave he had been initiated into a sect of high-caste natives, or Rishis as he called them, who studied the occult sciences.

I asked what this sect was. Was it a sort of native Free Masonry?

Yank laughed long and loud at this question. "Free Masonry!" he exclaimed in a tone of deep scorn. "That's a sort of parlour version of the real thing – and a fake antique into the bargain!" He said that the sect into which he had been initiated dated back thousands of years before Free Masonry. "Free Masonry," he exclaimed again with another scornful laugh, "that's all right for managers of pop-factories and would-be officers and gentlemen. I much prefer the poor old Buffaloes, from what I have seen and heard of them. The Buffs are a simple boozing benevolent club and make no pretence at mysticism."

My question seemed to have got his goat a little, so he told me more than what he otherwise might have done. He confided in me that he had now passed through the first and second circles of his sect and that what he had seen in the process was enough to drive ninety-nine men out of a hundred insane. He also said that it would take him some years of deep meditation before he would be able to pass through the third and final circle, which he gathered was far more of an ordeal than the other two put together.

What he said was Greek to me and I really thought he was going up the loop.

He seemed to read my thoughts, for he said: "No, Dick, I am not going up the loop, as you call it, but there are more things in Heaven and on earth than ever philosophers dream of. Although I have taken on to complete my twelve years with the Colours, I shall never complete them, and one day you will hear news of me which should not astonish you, although I dare say it will astonish others." He was extremely exalted on this occasion and tried to impress me further by saying, "You will know that I have gone to the mountain: taking my servant with me, to the well of testing and the fire of pardon and that I shall be living with my companions upon the earth and yet not on it, and we shall live there fortified without fortifications, and we shall possess nothing though having the riches of all the land."

This was worse Greek to me than the other and sounded almost like a translation of some bit of secret native ritual which he ought not rightly to have revealed; but I recalled it in the very words he used, some years later when I was back in civil life. For I heard news of him then which, as he had said, did not astonish me. After completing nine years with the Colours, six of them abroad, he made an application to be sent to the home-establishment, which was granted. He left the Battalion with a party of time-expired men, but disappeared after arriving at Mandalay. A few days later Lewis, his principal medium, also disappeared from the Battalion. After the usual number of days had elapsed they were posted as deserters, but although the

242

whole of Bunnah was combed, no trace could be found of either of them. They had vanished as mysteriously as if the ground had opened and swallowed them up, and from that day to this nothing has been seen or heard of them. I consider this most remarkable, for anyone would think that in a country like Burmah two missing white men would be bound to be seen by someone or other and gossiped about.

I have often asked myself the question: "Why did Yank make an application to go to the home-establishment when he knew full well that he never intended going there?" One answer would be that he still had a spark of vanity left in him and that deserting from the Army on his way to England would seem more mysterious than if he deserted while still having to serve some years abroad. But it is also possible that Yank had reckoned that the time when he would be ready to enter on his new life would come at about the same time as he would be sent down the river, and he would thus have a free trip to Mandalay, near where his Rishi pals seem to have had their headquarters.

As for the mysterious words he spoke to me, a friend of mine who is well versed in these matters tells me that the Rishis, who are a remarkable set of magicians and claim to have originally come from Ethiopia, are gifted with extraordinary powers, including second-sight and the power too, it is said, to kill or heal at a distance by mere triumph of mind over matter. They are also supposed to spend a lot of their time levitated three feet above the ground in honour of their gods; which may have been what Yank meant by "upon the earth and yet not on it." And they are supposed to make a sort of protective vapour to bivouac under, out of sight of ordinary eyes: which is what he may have meant by being fortified without fortifications. As for "possessing nothing, though having the riches of all the land," these Rishis are said to be well supplied with all manner of food, drink and other comforts by the superstitious natives in their neighbourhood. The well of testing and the fire of pardon are holy places on the secret mountain where the Rishis live. No white man knows exactly where this is. It is somewhere deep in the jungle. Each Rishi has a servant with him, whom he sends on messages to the outside world, and in order to ensure that the sacred secrets will be well kept, this servant has to be kept in a permanent condition of trance. Lewis would have come in very handy in this respect. For myself, I think that most, if not all, of this yarn is nonsense, but India and Burmah are funny places and there is no doubt but what Yank had strange powers which would be a good recommendation to the queer sect in which he claimed to have enlisted.

During the winter it was not too warm by day and not too cold by night; during the summer the temperature rarely went above 112 in the shade. Punkahs were then in use, but tatties were not needed. Chickens, ducks, eggs and fruit were cheap, if not cheaper than in India. A large full-grown pineapple only cost two pice; beef was dearer, but it was of better quality, especially at Rangoon, where I paid ten annas a pound for the best beef-steak. During the rainy season there was an abundance of rain which formed large and small lakes in different places. In less than a week after they were formed we used to go fishing in them. I don't know where the fish came from, or what became of those that were not captured when the lakes dried up-which most of them did during the winter. In one large lake, not far from Barracks, fish three and four pounds in weight were sometimes caught a few weeks after the lake had formed. There were a number of toddy-plantations around these parts and when the trees were tapped they produced a liquid not unlike cocoanut-milk, only a little thicker. I only tasted it once, but did not care for the stuff. It was said that a small glass of toddy, taken early in the morning, would do a man good. This may have been true, but I do know that men who drank a quantity of it during the afternoon generally regretted it the next morning. Just after dinner one afternoon four men I knew went to a toddy-plantation, with a quart-basin and eight annas between them. They were too broke for the Canteen but knew that the toddy-tree wallah would supply them with all the toddy they could drink for this small sum of money. They were fighting drunk when they arrived back in Barracks and had to be frog-marched to the Guard-room. One of them afterwards told me that he had woken up with the greatest fat head he had ever had in his life, and after drinking a pint of water he felt half-drunk again, the water starting the action of the toddy in his stomach again.

I visited Mandalay several times, but much preferred Rangoon. Mandalay, which our First Battalion helped to capture on November 28th, 1885, is a place of many Pagodas and outside the town is a large Burmese fort with the carved palace of King Theebaw inside it. After knocking around the palaces of the Mogul Emperors, I found this carved wooden palace rather commonplace. Still living in one wing of the palace was a very old man who had been Theebaw's chief minister at the time of the Burmese War. The Burmese believed that he had been granted a pension by the British Government as a reward for letting the British troops into the Fort when it was captured. The fort was a very large place with a high wall and moat around it, and now housed a battalion of native troops. Lulu's Bar was the most famous resort for white troops at Mandalay. Lulu was a half-caste woman. She kept a troupe of nautch-girls who were also prostitutes as well. For a little buckshee these girls would dance as naked as they were born. Lulu did a good trade on the side with bottled beer and strong drink. I was not overstruck with the nautch, a sinuous type of dancing which Europeans are as a rule unable to appreciate as much as the natives. The chief charm and point of the nautch is supposed to lie in the girl's manipulation of her fingers in various set movements that have some religious meaning or other, and also have special names, the same as foot-movements have names in European dancing. Many of the most celebrated performers are by no means good-looking, because looks are not so important to the native audience as that the fingers should be properly used.

There were only four British infantry battalions stationed in

Burmah: the Royal Scots Fusiliers at Rangoon, the Essex at Thayetmyo, the East Yorks at Maymyo, and ourselves at Shwebo and Bhamo. The East Yorks found a detachment at Mandalay who stayed in a small barracks on a hill two or three miles beyond the Fort. Maymyo, which is five or six thousand feet above sea-level, is built on a very large plateau and is connected by a narrow-gauge railway with Mandalay. I once spent a fortnight's leave at Maymyo attached to the East Yorks, whose regimental pet at this time was a magnificent cheetah which they first had when it was a puppy. It was a bit hard on the battalion's dogs. The bungalows at this place were only slightly raised off the ground and it constantly rained during the time I was there.

I was surprised at the number of Japanese and Chinese, especially the latter, that were living in Burmah, and also at the number of yellow-robed Buddhist monks and poongyees (priests) that I came across. Like the priests of all denominations throughout the civilized world, they looked very well fed. Just behind the Regimental Bazaar at Shwebo were a collection of primitive native huts in which lived Burmese, Hindoos and two Pathans. One of the Pathans had a young wife who had got very sweet on a Burmese lad of fifteen years of age. While her husband was out working during the day the boy frequently visited her. Someone must have given the husband the tip because he returned unexpectedly to his hut, early one afternoon, at the moment that the guilty pair were having connections. He seized hold of the boy and threw him out of the hut as easily as if he had thrown a bundle of rags. He then caught up his tulwar and sliced off three parts of his wife's nose, saying to her as he did so: "For the rest of your days you will look so repulsive that even a pig would think twice before having connections with you." He then gave himself up to the native police and told them that if the Burman had been a man he would have killed him on the spot, but it was not the Pathan way to take vengeance on children. For the slicing job I believe he was sentenced to two years' imprisonment.

I saw more vultures at this place than anywhere I had been to.

If a dead horse or cow was lying about these loathsome-looking birds would appear mysteriously from somewhere and would not leave the carcase until the bones were picked clean. And I often saw the queer-looking birds called Adjutant birds, which were also great flesh-eaters. They had very long legs, on which was perched a body similar to a duck's, only much larger. Their bills, which were thick and heavy, were very nearly two feet in length. To see one of these birds walking over the ground was not unlike seeing a long-legged adjutant of an infantry battalion walking over the parade ground. They had to run twenty or thirty yards along the ground, like an aeroplane, before they could rise in the air, and then they flew straight up into the heavens until they were out of sight.

That winter a two days' race-meeting was held at Shwebo. This was a flat and jumping meeting combined. The two principal races were the Shwebo Derby and the Shwebo Grand National. European bookies from Mandalay attended this meeting, which was well patronized by the Burmese, Chinese and other nationalities living around the place. They enjoyed a race-meeting as well as we did, especially the Burmese, who are, I dare say, the greatest race of gamblers in the world. At the time this race-meeting was held the Burmese were having a pooay, a festival which lasted seven days and was entirely devoted to gambling and enjoyment. Not far from the Barracks was erected an open air theatre where plays were given in the evening. Peep-shows were another form of amusement, and quite a number of wooden shacks and tents were also erected for gambling purposes. All the gambling games were similar to our own, but there was one I had not seen before: the requisites for it being an airgun, a dart-board and a feathered dart. This game was played in a very long tent which accommodated fifty and sixty players and the owner charged four annas for admittance. There was a pathway down the centre of the tent, which had on each side of it a raised board flooring. On this the players sat in groups of five, one man acting as banker to the other four. Just inside the tent stood a man with the airgun, who fired the small feathered dart. Opposite him, on the other side of the tent, at the end of the pathway, stood a man beside a dart-board that was hung on a nail driven into a post. After all the backers had backed their fancies, which was either an odd or even number, the man at the dart-board would give it a sharp spin. It spun around so fast that it looked like a blur, and it was then that the man fired his dart at it. All eyes were now fixed on the dart-board until it stopped spinning, when they could see for themselves whether the dart was fixed in an odd or even number. Before the man at the dart-board could shout the number the bankers were already paying out the winners and in no time the dart-board would be spun again. If the dart had stuck outside the rim of the numbers another spin was given to the dart-board and bankers could double their stakes if they liked. This was the fairest gambling game I have ever seen. There was no skill about it: if there had been, it would not have been a pure gamble, and it was an even money chance between bankers and backers. This tent was always full of players and if one only left it for a few minutes he would have to pay four annas again for admittance. The owner did not gamble on the game himself, nor would he allow his assistants to do so.

There were Indian fakirs travelling around Burmah. One of them that appeared outside our bungalow at Shwebo was the cleverest I have ever seen. After he had finished his ordinary performance he was asked the usual question – whether he could perform the Rope Trick. He replied that he could not, but for an extra collection of five rupees he would perform the Indian Duck Trick for us, which was just as marvellous. None of us had seen this trick before, so the five rupees were quickly collected and handed to him. I had heard and read a lot about this trick, but up to now had not dropped across a fakir that could perform it. He produced a thin brass chatty that would hold about three pints, which he filled with drinking water from the tank underneath the bungalow. After placing the chatty on three stones raised about a foot from the ground, he threw a handful of sandy soil from the ground into it. He now placed on the water a carved wooden toy duck, which a boy could have easily hidden in his clenched fist. He then retraced his steps to his assistant, who apparently did not help him in any way with this trick. He sat on his haunches a good twelve feet from the chatty and for a couple of minutes played some weird music on a reed instrument. I found it impossible to keep an eye both on him and on the duck, so focused my gaze on the duck. Suddenly he ceased playing and said: "Sahibs, big storm in chatty coming." He now began a soft chant and we that were watching the chatty rubbed our eyes in amazement, as we saw the water disturbed by angry waves which tossed the toy duck about like a ship on the ocean. This storm lasted for about half a minute and none of us could explain afterwards why the toy duck was not tossed out of the chatty. When the water became calm again it was still floating in the centre of it. The fakir now played another tune on his instrument, but not for quite so long as on the first occasion. After he had ceased playing he now said: "Sahibs, duck now plenty fish." Just after he had begun chanting softly again the toy duck dived in the water until only the tip of its rump was showing. Back it came to the surface and floated around for a few seconds before it dived again, which it did at least half a dozen times. Watching this toy duck fishing was exactly the same as watching a live duck fishing on a pond at home. The fakir concluded the performance by telling it to salaam three times to the Sahibs, which it did by rising and dipping its beak in the water.

About twelve months previously I had read in an Indian newspaper called *The Pioneer* two theories as to how this trick was performed. They were written by a man who claimed to have seen the duck trick performed on several occasions. One of his theories was that the fakir mesmerized his audience into believing that they saw the duck fishing and salaaming, when in reality it was stationary on the water all the time. He did not mention anything about the storm in the chatty, so I take it that the fakirs he had seen were of an inferior class to the one we had seen perform the trick. His other theory, which contradicted the first, was that the fakir operated the duck with his big toe on which was tied a single long hair, with the other end of it tied on the duck. This hair was made of three or four horse hairs spliced together, which lying on the ground would not be visible to people standing up who were watching the trick. There were about thirty of us watching this performing duck, but not one of us would admit that we had been mesmerized into seeing what we did. The only man who believed that we had was Yank, especially after I had related to him the storm in the chatty. It was only natural that being a hypnotist himself he should believe such a theory; but he had never seen the trick performed and was not present to witness it on this occasion. The horse-hair theory we proved to be nonsense: before the fakir commenced to play his instrument a few of us brought our pocket knives into action. We cut through the ground between the fakir and the duck for a distance of about six feet in front of where we were standing; if an horse hair or anything else had been running along the ground our knives would have been bound to sever it. I have met many old soldiers of different units since then who had seen the trick performed at one time or another, but not one of them had seen the storm in the chatty, although several had heard talk of it.

Late one evening in a tent I watched a Burman, who was dressed in a fine silken dress, having a gamble on a game similar to Crown and Anchor; the only difference being that the figures on the dice and board were the Ace, King, Queen, Jack, ten and nine. He had taken a fancy to the King, on which he had doubled up until he was broke. He then asked the banker, who was a Chinaman, if he would lay him a certain sum of money against his silken dress, which he would bet on the King at the next shake of the dice. After examining the dress, the Chink agreed to lay the sum mentioned, but only on condition that the Burman undressed himself and laid the dress on the board before he shook the dice. It did not take the Burman a minute to slip behind a tree and undress himself. When he came back to the board he was clad only in a loin cloth and was as cool as a cucumber as he laid his dress on the King. The Chink after giving an extra shake to the dice brought the cup down with a bang on the board. When he lifted it the King was not showing, and the Burman in addition to losing his money had lost his best Sunday suit as well. This did not seem to worry him in the least. He watched the Chink shake the dice and lift the cup a few more times without the King showing, and then walked away with a grin on his face as much as to say that it might have been worse.

HOME AGAIN

Although there were supposed to be small bands of Dacoits still in Upper Burmah no precautions were taken for the safety of the rifles, such as had been taken in India. There were no heavy rifle racks with locks and chains in the bungalows, but through force of habit most men still locked the bolts of their rifles in their boxes over night. On manoeuvres, which we did around Amphora Shore, when we slept in the open most of us still slept with one leg through the sling of the rifle. This was not considered necessary in Burmah, but we could not break ourselves of the custom. The authorities evidently did not fear an uprising on a Sunday in Burmah, as no rifles were carried to Church; writing from memory, I believe the dress for Church parade was a belt and bayonet. When the Dacoits did put in an appearance they came looking for cash, not rifles. The Army Temperance Room at Shwebo was only slightly raised above the ground. The Secretary and Barman who slept in a bunk close to the bar woke up one night to find a Dacoit standing over each of them with a dagger in his hand. A few seconds later another Dacoit who had been overhauling the bunk discovered the box which held a week's takings, amounting to about three hundred rupees. Without a word one of them blew out the lamp, which was dimly burning, and disappeared in the darkness outside. The Secretary and Barman had been powerless, and by the time they had jumped out of bed and rushed into the open with their bayonets in their hands the Dacoits were far away. The only clue they had left behind was a dagger which one of them had dropped about six yards from the bunk. This was found at daybreak, but proved of no value, except as establishing the Secretary and Barman's innocence, in case that anyone should have suspected that the Dacoits were only a manner of speaking.

At various points along the river north of Mandalay were small cemeteries, in which were buried the men who had fallen in the Upper Burmese War of 1885. The graves had been allowed to get into a deplorable condition, but now through an appeal from the Colonel every man in the Battalion willingly gave a day's pay towards renovating these graves; the officers also made a handsome subscription. The graves of the men of other regiments besides our own were also seen to, vegetation was cleared, new crosses erected and in each small cemetery a large cross was fixed with all the names of the occupants on it.

The two largest pythons I have ever seen were shot outside an old ruined Pagoda, about a mile and a half from the Signallers' bungalow at Shwebo. One measured over twenty feet and the other just under twenty feet in length. They were shot by a half-caste who had gone out for an afternoon's shooting. He believed that the old Pagoda had been their home and said that they were mating when he shot them. Mating pythons are a very rare and very strange sight. They stand upright, propped on their tails, and coil around each other like a shining twisted pillar, squirming and hissing: they make an easy mark for a bullet when in this position. Only that morning I and a signaller named Sadler had been sent out to the old Pagoda, which was built on a slight rise in the ground. For the best part of three hours we were just outside it, sending heliograph messages to our young signallers, who were in pairs, a short distance apart, near the bungalow. It was not the first time we had been sent to the Pagoda for this work but we had never spotted the pythons; if we had done so, the revolver that the Prayer-wallah had given me would have come in very handy. But if they had spotted us first, and stalked us, once they had wrapped their great coils around us from behind while we were busy at our job, our numbers would have been up; and in due time, no doubt, we should have been posted as deserters from His Majesty's Forces. This half-caste who shot the pythons was a clever hunter. I met him once with a litter of four or five cheetah-pups, only a few days old. He had just shot the mother of them outside their lair. He told me that this was the third litter of cheetah-pups he had got since he started being a huntsman, but every pup of the previous two lots had died before it was a month old. He was going to feed this lot on goat's milk from a baby's feeding bottle.

Ever since I had been abroad I had kept up a regular correspondence with my Aunt. Three months after I arrived in India my Uncle had gone out to Italy to work in a new tinplate works that had been erected somewhere in that country. Twelve months later I received a newspaper cutting from the *Western Mail* which gave me further news of him – in spite of being well past middle age he had created a world's record in the number of boxes he had rolled in an eight hours' shift. But my Aunt knew nothing about it until she read it in the same paper. His record, however, was broken some time later by a younger man, a roller from Llanelly, Carmarthenshire, who was working in the same place. All my Uncle's letters home were written by some friend there. When he had worked in Italy for six years he returned home for good. When he got back I had only a couple of months to serve before becoming a time-expired man. After some argument with myself I decided not to take on to complete my twelve years with the Colours or to take my discharge in the country. The majority of my best pals had already gone and the rest would be going soon. Also I wanted to see Elaina and my Uncle and Aunt and cousins again. I would let matters take their own course. And one good thing was that when I went I would go straight home, not to the camp at Deolalie, thus avoiding the danger of contracting the Doolally Tap. During the last six months of my soldiering I had been on the steady, to save a few corks to carry back with me. My only expenditure on luxuries had been four annas a week for tobacco and four annas a day for beer. A few days before I left the Battalion I changed my rupees into sovereigns with a native shopkeeper in the Regimental Bazaar.

This old shark charged me fifteen rupees, four annas, for each sovereign, but if I had wanted to change one or a hundred of them into rupees he would have given me a bare fifteen rupees for each of them.

About the latter end of February, 1909, a draft of about sixty of us time-expired men returned our rifles to store, paraded for the last time and left the Battalion to proceed on our journey home. We were all mighty glad when we heard that we were proceeding by train to Rangoon and not being put on one of the river flats. We broke our journey to stay a night in the barracks beyond Mandalay Fort. Plague was raging in Mandalay, which was out of bounds for British troops by the time we arrived there. On arrival at Rangoon we boarded a Royal Indian Marine steamer which had, aboard her, time-expired men from the other units stationed in Burmah. Also aboard were some invalids and mental cases who were being sent home. It was a ten days' voyage to Bombay, with one call at Madras to fetch off some more time-expired men. The most interesting man in the ship was one of the mental cases, whom we soon nicknamed "the Champ." He was a full corporal of the Essex Regiment and a middle-weight boxer of repute, who had been sane enough until Jack Johnson defeated Tommy Burns for the Heavy-Weight Boxing Championship of the world. The thought of a black man being the World's Boxing Champion had so preyed on his mind that he had fairly gone up the loop. He was firmly convinced in his own mind that he was the only white man in the world who was capable of defeating Jack Johnson and regaining the championship for the white race. He had spent his last three weeks ashore in a padded cell in hospital and was placed in a padded cell of his own when he came aboard. As soon as he arrived on board he believed that he had been matched to fight Jack Johnson and that he now had to go into strict training for the battle. From the time he got up in the morning until he retired for he night he was in constant training. He did all the exercises known to man, and if his padded cell had not been a strong one he would have soon

255

punched a hole through it. He was only taken to the upper deck when it was necessary for him to go there, but was never violent with his escort who, he thought, had been sent to guard him from foul play during his training. Every morning he was taken out of his cell for about an hour's walk up and down the lower deck. After pacing the deck for a few minutes he would suddenly stop and do a few minutes' shadow boxing. His escort, who humoured him in every way, wisely stood aside while he was doing this. He always finished this shadow boxing with a terrific blow, saying as he did so: "You black bastard, when you stop one of these you'll be out for more than ten seconds." Our rations, which included bread, were very good on this ship, but with the training he was doing they were not sufficient for him. Any food we had to spare was taken to him and the men of his own battalion aboard took him many luxuries as well. He ate like a horse and looked in magnificent physical condition, and every morning the Doctor had a look at him through the spy-hole of the cell to see how he was getting on. Only once during the voyage did the Doctor enter the cell to have a chat with him. The Champ was quiet enough that morning and nobody would have thought he was stone balmy. After conversing normally for a few minutes he suddenly aimed a terrific uppercut at the Doctor's chin, saying as he did so: "Here's a blow I've been practising for days; what do you think of it?" The Doctor, who had been keeping a wary eye on him, dodged the blow by falling backwards out of the open cell-door into the arms of one of the Champ's escort. He afterwards said that if the uppercut had connected with his chin it would have stopped his thinking powers for ever.

When we arrived at Bombay we were transhipped to a larger trooper, where the Champ was issued with a suit of black civilian clothes, which he wore on the voyage home. I was told that all mental cases proceeding home from Bombay were issued with similar suits. After he had changed into his civilian suit a man of the South Staffords, who had been doing turns on the stage, gave him a box hat to wear with the suit. The box hat fitted him perfectly and when he looked into a mirror and saw the way he was dressed his madness took another turn. He now believed that he had already fought and knocked Jack Johnson out and was now Heavy-Weight Boxing Champion of the World. He never did any more training, but whenever he was offered a cigar to smoke he would give a detailed account of the fight and a demonstration of the blow that had knocked the Negro out. We were all hoping that he would regain his reason during the voyage home, but he did not. When he arrived at Southampton he still believed that he was the Champion and was sent to another military asylum.

From the time I left Rangoon I kept sternly away from the Crown and Anchor board, contenting myself with the game of House. At Aden we took aboard some more time-expired men and also half a dozen mental cases, two of them being pretty hopeless cases through self-abuse. Aden is known as one of the most dead-alive and dreary stations that troops can be sent to: it has great strategic value, that is all.

After being thirty-one days at sea, uneventful ones too, I was pretty glad when I landed at Southampton, from where time-expired men were sent to Fort Brockhurst, Gosport. Here I was issued with a cheap ready-made suit, which fitted me fairly well, and then despatched to my home, being now transferred to the Army Reserve. But first we said a lingering goodbye to one another over our mugs of neck-oil in the Canteen, and it was quite a queer experience being all in civilian clothes. When we first gathered around the bar we had a job to recognize one another: a man of my company remarked to me that it was like a couple of caterpillars that have been bosom pals all their life, nibbling away at the same cabbage-leaf, day in, day out, and suddenly they meet again as moths or butterflies and begin to address each other as "Mr." instead of Jack and Dick.

When I enlisted I had no distinguishing marks or scars on me but I was now tattooed with designs of animals, snakes and celebrities on my arms and chest. It was considered fashionable for a soldier to have some design or other tattooed on him, and most of us had it done to us during my time. I have always regretted being tattooed and some men I knew must have regretted it more than me. Some of them had tattooed on their backs a pack of hounds in full cry after a fox, with the fox seeking cover in the hole of the backside. A few had "A Merry Xmas" tattooed on one cheek of their backside, with "A Happy New Year" on the other. I drew the line at designs of this sort and was very glad I did so, especially when I went to work in the collieries, as it was customary for men returning from work to have a bath in a tub in the kitchen of the house. At one place where I was lodging there were generally one or two women in the house having a cant with the landlady while I was having a bath. There was nothing in this, because when a man slipped his trousers off to wash his legs he covered his nakedness with one hand and washed himself with the other. But if one of those designs had been tattooed on me, every woman in the street and neighbouring streets whom the landlady was friendly with would have come visiting. They would have popped into the kitchen to see her on some pretext or other, judging the correct time that I was naked in the tub.

My Uncle and Aunt, who were now living at Liswerry, Newport, hardly knew me when I walked into the house. During my service I had grown three inches in height and gained over three stone in weight. My Uncle and I celebrated my return by having a jollification that evening, and during the next few days I visited other relatives at different places in the neighbourhood. I found it delightful for a time to see green fields and meadows with English flowers and hear the birds singing: it was something I had not seen or heard for six years and a half. It was also strange at first to see so many white people about. With their white faces most of them looked to me as if they were ill. I must have got this impression by seeing during my time abroad only British soldiers, Government officials and a few tough planters. The English meat, bacon and other foods tasted delicious after what I had been accustomed to abroad – I honestly believe that there is more nutriment in one pound of good English steak than there is in a whole cow in some parts of India. And after being addressed as Dick for so many years I found it peculiar, for a time, to be addressed by my own Christian name. Frank and Dick seemed quite separate persons.

First I worked at odd jobs here and there around Newport, and then my Uncle, who had lately started to work in Caerleon tinworks, about three miles from Newport, got me a permanent job there. After my hands had hardened, the work was easy enough, but I was not very happy. My delight with home wore off and I found myself longing to be back with the Battalion again in the Tropics. When taking a walk in the evening I missed the croaking of the bull-frogs and the buzzing noise made by the tropical insects. During the night I missed the howling of the jackals and the cry of the cheetah. Before I left the Battalion I could have taken on to complete my twelve years with the Colours, but now I was on Reserve I could not do that. I had soon written to the military authorities on the subject and received a reply that no reservists were allowed to rejoin the Colours. I could quite understand this, as the Government could always call upon the reservists when they were wanted, and it was far cheaper to pay them sixpence a day, which was paid in a lump sum at every quarter of the year, than to keep and pay a time-serving soldier. I seriously thought of joining the Army again as a recruit under an assumed name. I would have done so, but I hated the thought of again going through a recruit's training and of serving in any other regiment than my own. I knew several men who had done this, also men who had deserted from one regiment and joined another. They were always liable to be arrested for fraudulent enlistment or desertion, and they were the type of men who soon got fed up with any kind of life.

After working a few months I left my job and, taking my shaving-tackle and toothbrush with me, tramped the country for a couple of weeks with fifteen shillings in my pocket. I made believe that I was on the line of march from one station to the next. Late one evening I picked up with an old tramp who, by his conversation, seemed to be making for nowhere, the same as I was doing. The following morning he must have thought I was an escaped lunatic; for when I stopped by a stream to clean my teeth and have a wash and a shave he hurriedly left me. This was in August. Whenever I came to a river I indulged in a swim, and sometimes followed the river up to its source. After a week I had a washing day, washing my shirt and socks, and passed the rest of that day away watching them dry. My food was bread, cheese and onions, with an occasional pint of beer, which was only threepence a pint. My fifteen bob lasted me a fortnight and I still had a few coppers left when, feeling greatly benefited by my tramp, I drifted back to Blaina. I began work in one of the collieries there, where I continued until 1914. With the exception of the tin-works, which had been dismantled, every other industry at Blaina was still in a flourishing state. It is difficult to believe that such times existed, now that eighty per cent of the population of the village is permanently unemployed and has been so for many years. There is one problem that someone may be able to solve simply but seems very queer to me. In the prosperous times in Blaina there were two pawn-shops that did a flourishing trade, but no banks. Since 1921 the pawn-shops have gone out of business, but the branches of two powerful banks have taken their place and seem to do a fair business. But this is post-War talk and I must return to 1908.

I did not work at the coal-face, hewing coal, but was a timberman's assistant. Underground workers did an eight-hour shift with a twenty minutes' allowance for food, halfway. The wages of a timberman's assistant then was thirty-five shillings a week, and of a labourer thirty shillings a week. At the present time underground workers, when they happen to be in employment, work seven and half hours a shift, timberman's assistants and labourers being paid equal wages, which is about forty-five shillings for a full week's work. I worked on the night shift, when all the repairing work was done. The timbermen and their assistants were engaged in making height on the main roads, clearing falls of rubbish, and timbering the sides and holes overhead where the falls had occurred. Many a night the timberman I was working with had his work so well under hand that we could afford to spend more than an hour over our food. If there were other men or boys working near us who were in the same fortunate position we would collect together in one place to have our meal together. At these grub-times I was often called upon to relate my experiences abroad, which I did in much the same manner as the old road-man at the commencement of this story. "India," I would say, "is a land flowing with milk and honey. And what I was thinking of when I left it to come back home here and work again deep in the bowels of the earth, I'm damned if I know." The boys would fix their eyes on mine and drink it all in.

During the first year I was in England I did not have a pain or an ache and I thought I had got rid of my old enemy, malaria. The second year during the latter half of the summer, I discovered that it was still in my bones. I doctored myself with quinine and whiskey, but had to lose a week's work over it. Every other year after that, down to 1917, I got an attack which laid me up for four or five days. Since then I have no recurrence of it, for which I am extremely thankful. My Uncle died at the ripe age of seventy-five, and if I keep my health and spirits until that age, the same as he did, I shall consider myself a damned lucky man. My dear old Aunt is still alive and still believes that some day I shall see the error of my ways. She reads her Bible daily, but my reading is mostly confined to the forms and lists at the local Labour Exchange where I have a temporary job as a clerk. I hate pen-pushing, but the job was forced on me by a disability contracted during the War, which prevented me from continuing work down the pit. It has been with the greatest difficulty that I have struggled through these chapters, during periods of temporary unemployment and on Sundays and Bank holidays. If I had known what I was letting myself in for I should never have had the heart to commence. My feelings on laying down my pen will be very different from what the historian Gibbon says that he felt on the day that he completed his famous work, which we had in our Regimental Library at Agra, on *The Decline and Fall of the Roman Empire*. The historian Gibbon felt a mild melancholy steal over him, like parting from an old friend; I shall feel more like adjourning to the nearest beer-fountain for a quart or two of purge to celebrate the conclusion of a very severe sentence that I innocently passed on myself.

In 1912 I extended my service for another four years on the Reserve. I little thought when I did so that two years later I should be called back to the Colours to rejoin my old Battalion again. Every quarter-day, or pension-day as it was called, a number of us reservists and service-pension-wallahs would have a day off from our work to spend it together in the Castle Hotel. There were more natives and Chinks killed in one day at that hotel than there had ever been in the Indian Mutiny and Boxer Rising. If the old Bacon-wallah of Meerut had been in our company on a pension-day he would have felt kindly disposed to one man who, late in the evening, never failed to hang half a dozen natives on the gas-brackets after every pint he drunk. The Hotel had the reputation of having the best beer in the neighborhood, and we always voted it an excellent drop of neck-oil. By stop-tap most of us had said what utter fools we had been to leave the Service, and that if we had our time over again we would not leave the Army until we were damned well kicked out of it.

LIBRARY OF WALES

The Library of Wales is a Welsh Government project designed to ensure that all of the rich and extensive literature of Wales which has been written in English will now be made available to readers in and beyond Wales. Sustaining this wider literary heritage is understood by the Welsh Government to be a key component in creating and disseminating an ongoing sense of modern Welsh culture and history for the future Wales which is now emerging from contemporary society. Through these texts, until now unavailable or out-of-print or merely forgotten, the Library of Wales will bring back into play the voices and actions of the human experience that has made us, in all our complexity, a Welsh people.

The Library of Wales will include prose as well as poetry, essays as well as fiction, anthologies as well as memoirs, drama as well as journalism. It will complement the names and texts that are already in the public domain and seek to include the best of Welsh writing in English, as well as to showcase what has been unjustly neglected. No boundaries will limit the ambition of the Library of Wales to open up the borders that have denied some of our best writers a presence in a future Wales. The Library of Wales has been created with that Wales in mind: a young country not afraid to remember what it might yet become.

Dai Smith

LIBRARY of WALES
FUNDED BY

CYNGOR LLYFRAU CYMRU
WELSH BOOKS COUNCIL

PARTHIAN
A CARNIVAL OF VOICES
WWW. PARTHIANBOOKS.COM

WWW.THELIBRARYOFWALES.COM